Reflection in the Waves

Breakthroughs in Mimetic Theory

Edited by William A. Johnsen

Reflection in the Waves

The Interdividual Observer in a Quantum Mechanical World

Pablo Bandera

Michigan State University Press

East Lansing

♾ The paper used in this publication meets the minimum requirements
of ANSI/NISO Z39.48-1992 (R 1997) (Permanence of Paper).

Michigan State University Press
East Lansing, Michigan 48823-5245

Printed and bound in the United States of America.

28 27 26 25 24 23 22 21 20 19   1 2 3 4 5 6 7 8 9 10

Library of Congress Cataloging-in-Publication Data
Names: Bandera, Pablo, author.
Title: Reflection in the waves : the interdividual observer in a quantum
mechanical world / Pablo Bandera.
Description: East Lansing : Michigan State University Press, [2019]
| Series: Breakthroughs in mimetic theory | Includes bibliographical references and index.
Identifiers: LCCN 2018020112 | ISBN 9781611862829 (pbk. : alk. paper)
| ISBN 9781609175627 (pdf) | ISBN 9781628953299 (epub) | ISBN 9781628963298 (kindle)
Subjects: LCSH: Quantum theory—Philosophy.
Classification: LCC QC174.12 .B35525 2019 | DDC 530.1201—dc23
LC record available at https://lccn.loc.gov/2018020112

Cover and book design by Erin Kirk New
Composition by Charlie Sharp, Sharp Designs, East Lansing, Michigan
Cover art © Ali Mazraie Shadi. All rights reserved.

Michigan State University Press is a member of the Green Press Initiative and is
committed to developing and encouraging ecologically responsible publishing
practices. For more information about the Green Press Initiative and the use of
recycled paper in book publishing, please visit *www.greenpressinitiative.org*.

Visit Michigan State University Press at *www.msupress.org*

To my father,
without whom this book
would have been impossible.

To my wife,
without whom everything else
would be unthinkable.

Contents

Introduction

The subtitle of this monograph is intended to give two general impressions: that it has something to do with modern physics and something to do with René Girard's theory of mimetic desire. That these two subjects have anything to do with each other is admittedly a strange claim to make, and I should probably apologize for asking the reader to take his or her first step in this study over a potential stumbling block. Then again, readers of modern physics are used to odd contrasts and apparent contradictions, and readers of mimetic theory are used to stumbling blocks. So it is precisely these readers who should be most open to an underlying synergy between the two fields of study, and it is this synergy that I wish to explore.

The greater resistance will come, of course, from the

scientific community, which has acquired a kind of allergy to philosophical questions. It has acquired this allergy, however, not from any inherent incompatibility with philosophy, but as a reaction to the persistent and increasing relevance of philosophical questions that keep imposing themselves on scientists. The enormous amount of literature relating science and philosophy, which has been steadily growing in recent years, testifies to this basic synergy. And yet the prevailing attitude of many modern scientists toward philosophy, and especially toward any reference to the human or to the divine, is decidedly negative. An allergic reaction becomes more acute, more violent, as it gets more surrounded by the allergen. This, I believe, is essentially what we have been witnessing for the last hundred years or so, with occasional uncontrollable fits such as Stephen Hawking's proclamation that "philosophy is dead."[1]

Nonetheless, as science has become more "modern," the range and depth of philosophical questions have increased. For example, as soon as one enters into a discussion of quantum theory and its strange implications, one feels compelled to answer a fundamentally philosophical question: Are we dealing here with a problem of ontology or a problem of epistemology? Do the strange implications of quantum theory say something about the world itself or merely what we *know* about the world?

There is a temptation to assume the latter and console ourselves with the idea that the universe is not *really* arbitrary or irrational, only mysterious and elusive. This has in fact been the explicit position of many physicists, beginning with Werner Heisenberg. But even if we speak of quantum theory as describing only what we know about the observed physical system, this is not intended to mean what we are capable of knowing today, with our present technology or with the current limitations of our brains. It is what we can know *in principle*—what nature allows us to know, and therefore what is *inherently* knowable. The "uncertainty" quantified by Heisenberg is meant to describe the objective world, not the subjective observer. This is especially evident when one considers that Heisenberg's original term *Unbestimmtheit* does not primarily mean "uncertainty," which suggests a limitation or confusion of the mind, but is more correctly translated as "indeterminacy." Despite his appeal to epistemology, Heisenberg understood that, after all, it is not we who are uncertain but the world.

As tortured and confused as this philosophical question has been since Descartes, it has been the bane of modern physics since Schrödinger introduced his infamous cat.[2] At the heart of all the strangeness of quantum theory lies a struggle to understand the connection between the subject and the object of an observation—between the observer and the observed. This is, of course, one of those big questions that have been discussed and debated in philosophical circles for many years, with implications well beyond the scope of this short monograph. Nonetheless, in my brief study I will need to address this and a few other big questions of science and philosophy, but only just enough to clear the path forward as we follow the light that quantum theory and mimetic theory shed on each other.

For example, the tug-of-war between subject and object has been fueled by another debate that shows no sign of abating anytime soon. Modern physicists agree on the critical role of the observer but are starkly divided on the question of who or what this observer is. The word "observer" naturally brings to mind the image of a human person, with eyes that see and ears that hear. But while some have argued for the necessary humanity, or at least consciousness, of the observer, most modern physicists reject the notion that we humans are special in any physically significant way. As a method of enquiry or a path to genuine understanding, this debate has been woefully disappointing. We have already seen it run repeatedly to its extremes: on the one hand a sterile reductionism, on the other a vague monism or poorly veiled pantheism. The present study will try to take both physics and the human person more seriously than that.

There is another topic that has played a smaller role in the philosophy of science but is of central importance for our purposes here. This is the intriguing similarity between the probabilistic wave function in quantum theory and the Aristotelian concept of *potentia*. These parallels were first noticed by Werner Heisenberg and have been elaborated in different ways by a number of other physicists and philosophers. Most approach the subject from the perspective of a generic observer—that is, without considering any special qualities of the specifically human observer. But I believe the parallels between probability and potentia are much more profound and revealing when the latter is understood, not merely in Aristotelian terms, but in the fuller sense developed by Thomas Aquinas. To do this, we must indeed take into account certain unique qualities of the human observer.

All of these big questions can be tied together in a coherent and powerful way when one considers an insight that is in fact *not* discussed or debated in any scientific literature at all. I am speaking of the common assumption, taken for granted by both classical and modern science, that the act of observation is characterized by a direct independent relationship between the observer and the object of observation. That is, despite the complexities developed over the last century around the nature of observation and the "measurement problem," we still ultimately rely on the rather intuitive notion that "my observation" is indeed "mine," independent of anyone else's, and relates me directly to the thing I'm observing. All models of observation assume this fundamental connection between me and my object and then describe how other objects and observers are or aren't related to it. As we will see, it is precisely this simplistic assumption that quantum theory calls into question, and this will force us to reevaluate what it means to be an observer and what it means to be observed. But the point is not that quantum theory is wrong, much less that quantum mechanics is somehow not real, or not really "quantum mechanical." The point is that quantum mechanics is trying to tell us something, and our rather limited understanding of the nature of observation has prevented us from seeing what that something is.

The philosophical problems of quantum theory are considered problems of the modern world and a product of the twentieth century. But their roots stretch back much further, at least to the thirteenth century, and really all the way back to the first century AD. But we should take things one step at a time and start from the beginning, which in this case is toward the end, when the line between classical and modern physics was being drawn in bold strokes. It was at this recent point in history when we first noticed the beacon being signaled from the unreachable depths of nature—a beacon of warning, but also ultimately of hope.

1

The Collapse of the Rational World

Toward the end of the nineteenth century, the scientific community was at a loss to explain the results of an experiment that, by all accounts, should have been fairly straightforward. The experiment involved the measurement of thermal energy emitted from a purely radiative source—what is known as "blackbody radiation"—and plotting the intensity of this energy versus its wavelength. The expected results were easily calculated using well-established equations of electromagnetics and thermodynamics. According to these fundamental laws, the measured radiant intensity should have increased monotonically with decreasing wavelength, generating a curve that swept consistently upward toward the top of the graph. But when the experiment was actually conducted and the results plotted, a very different and disturbing picture was formed. The curve started out well enough, climbing faithfully according to expectations. But then, as it entered the shorter wavelengths of ultraviolet light, the curve, in utter rebellion against the established laws of physics, turned downward and decreased swiftly to zero. There was no explanation for it. It was impossible. And yet the results were stubbornly repeatable, and the truth of it (whatever that truth was) was unavoidable. This seemingly simple blackbody was telling us something—something like, "Welcome to the twentieth century; things are going to be different around here."

It came to be known as the "ultraviolet catastrophe" because it was in this region of the blackbody radiation plot that our understanding of the world proved to be fundamentally wrong. There was nothing to be done about the experimental results themselves. So instead, the German physicist Max Planck reformulated the old equations around a crucial and original hypothesis: perhaps the atoms of a blackbody do not absorb or emit energy continuously, as we thought, but do so only in discrete amounts, which Planck called "quanta." It was a guess, with no scientific precedent, but it made the mathematics match experimental results perfectly.

When Planck presented his results to the scientific community in 1900, it was not obvious to anyone (including Planck) that this was anything other than an ad hoc solution to a particular mathematical problem. But a few years later, in the context of yet another experiment that showed a similar disregard for the laws of physics, Albert Einstein applied Planck's hypothesis to light itself (as opposed to the atoms of a blackbody), postulating the existence of quanta of light which he called "photons."[1] Suddenly, this notion of "quantization" became, not just an isolated peculiarity of blackbodies, but a property of one of the fundamental building blocks of nature. This powerful new idea attracted and inspired the best scientific minds of the time, so that by the third decade of the new century the distinction was already made between "classical physics" and "modern physics."

Just as classical physics has its fundamental equations (Newton's laws of motion, Maxwell's equations of electromagnetism, etc.), the new physics has its equation as well. This is Schrödinger's equation, and it is, oddly enough, a wave equation. That is, the solution to Schrodinger's equation, called the "wave function" or "state function," is not a single number such as 10 meters of length or 25 degrees centigrade. It is a wave-shaped distribution over a range of numbers, as shown in figure 1. The equation works very well and can be used to predict a wide variety of physical phenomena with uncanny accuracy. So it seems we have the right equation. The question is, what does this equation mean? If we use Schrödinger's equation to calculate, say, the position of a particle within a certain volume of space, what does it mean to say that the particle's position is distributed over this volume? Intuitively we would like to point to the particle and say, "It is there." The new physics seems to be telling us we can't point anymore, we can only wave our hands. In fact, in quantum theory we do not generally speak of the position of a particle. Instead we say that the particle is in "superposition," referring to this range of physical states contained within the wave function that together make up the overall state of the particle.

In 1928 Max Born came up with an explanation that ultimately formed the foundation of our new understanding

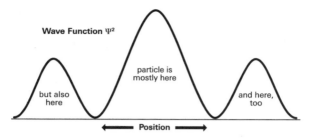

Figure 1. A Simple Example of the Quantum Wave Function

of the world. He suggested that the solution to Schrödinger's equation is actually a *probability* distribution. It represents the probability that an object or system is in a particular state at a given time. In the case of our particle, when we calculate its position, the wave function doesn't tell us, "It is there," but rather, "It is *probably* there."

This was a neat and well-defined answer to the question at hand, but one with profound and disturbing implications, especially when one considers the scientific notion of completeness. In general, the laws of physics are intended to provide a "complete" description of a certain object or system. For example, if we use Newton's laws of motion to calculate the vertical position of an object in the air, we will get an answer like "5,000 feet above the ground." This is a complete description of the vertical position (or altitude) of that object. It is not 4,000 feet above the ground, or 6,000 feet, but 5,000 feet. There is no more to say about the altitude of this object. Similarly, Schrödinger's equation is intended to be a complete description of the state of a system. It is not that the particle is really here, even though the wave function says it might also be over there. The wave function is not wrong or superfluous. In other words, we do not use Schrödinger's equation (or Newton's) to tell us where the particle *isn't*, we use it to tell us where the particle *is*. But if the wave function is a probability distribution, then where the particle *is* is where it *might be*. These two concepts—where something is and where it might be—are strangely confused together. There seems to be a fundamental ambiguity inherent in the very state of the particle, and in the state of any physical system.

This ambiguity was quantified by Werner Heisenberg in his famous uncertainty principle. By combining some of the new ideas developed by people like Einstein and Louis de Broglie, Heisenberg derived a set of mathematical relations that effectively imposed a hard physical limit on what was fundamentally knowable in nature. These relations have a simple form, for example:

$$\Delta x \cdot \Delta p \geq \hbar/2.$$

The symbol Δx is the uncertainty (or, as Bohr called it, the "latitude") in an object's position, and Δp is the uncertainty in that object's momentum (a similar relation involves energy and time). Heisenberg's uncertainty principle states that the more an observer knows about an object's position, the less he or she knows about its momentum, and vice versa. Moreover, this knowledge can never be absolute—the uncertainty in either one must be at least $\hbar/2$, which equals $5 \cdot 10^{-35}$ J·s. In other words, there is in the very heart of nature a minimum amount of uncertainty. We are not referring to the limitations of our measurement technology, but to what is measurable in principle. The universe is inherently indeterminate.

Werner Heisenberg and Niels Bohr were the primary champions of this new understanding of the universe. Their collaborative work in Copenhagen in the late 1920s became the philosophical foundation of quantum theory. The basic ideas have been developed and molded into several variations, but because of their origins they are rather broadly referred to as the Copenhagen school of thought. Despite its scandalous implications, it remains to this day the most widely accepted interpretation of quantum theory.

Many of the great pioneers of the new physics, including Schrödinger, Einstein, and de Broglie, rebelled strongly against this interpretation. Einstein's succinct response to the suggestion that the universe is indeterminate was "God does not play dice" (to which Bohr replied that Einstein should stop telling God what to do). There has been no shortage of memorable arguments and anecdotes over the last hundred years as physicists and philosophers have tried to make sense of the new science. But it was Schrödinger who produced the most enduring and well-known image of quantum theory—the image of a helpless cat in a box. In 1935 he proposed a thought experiment to show how this new understanding of atomic physics can lead to "quite ridiculous cases"[2] when applied to everyday macroscopic systems. This now-famous experiment goes as follows: There is a box, with no windows of any kind so that no one outside the box can see its contents once it is closed. Inside there is a small amount of radioactive material with a half-life of one hour. This means that after one hour there will be a 50 percent chance that the material will have decayed and emitted an atomic particle, which would then be detected by a nearby Geiger counter. The Geiger counter in turn is connected to a small hammer in such a way that, if a radio-active decay is detected, the hammer will come down onto a vial of cyanide, smashing it and releasing the deadly poison into the air. Finally, lying next to this contraption inside the box, blissfully ignorant of its impending doom, is an innocent cat. Let us now close the box and wait for precisely one hour. At that time, what can we say about the "state" of the cat? Is it alive or dead?

Most reasonable people would answer quite simply that there is a fifty-fifty chance that the cat is dead, since there is a fifty-fifty chance that a radioactive particle was emitted. Therefore, the cat is either alive or dead. A quantum physicist, however, would disagree. He would point out that the wave function describing the state of the cat is comprised of both possibilities (or probabilities) of live cat and dead cat, and this is a *complete* description of the cat's state at that moment in time. The cat, then, is not *either* alive *or* dead, it is somehow alive *and* dead simultaneously. It is this "blurred model"[3] of nature that Schrödinger did not trust.

But even our quantum physicist would have to agree that, when the box is opened and we look inside, we can only see either a dead cat or a live cat. Abstract speculations of some sort of hypostatic union within the mysterious confines of a system we cannot see may be acceptable in theory, but ultimately our own eyes cannot deceive us. Once the box is opened, the wave function of the cat can no longer be a combination of possibilities but must suddenly become either one thing or the other. This in quantum mechanics is called the "collapse" of the wave function, and it is perhaps the most scandalous element of quantum theory.

In classical physics, when two observers perform identical experiments under identical conditions, the results of both experiments are expected to be the same. The physical world does whatever it does according to the preset laws of nature, regardless of any observer that may or may not be watching, and this leads to predictable, or at least repeatable, results. If the two results are not the same, then at least one of them is simply wrong.

The disturbing suggestion of quantum theory is that, if two identical experiments yield different results, they may both in fact be correct. The wave function of a system reflects a combination of possible measurement results, and it collapses to only one of those possibilities the moment it is observed or measured. But the state to which the system collapses is not determined by any causal natural laws. It is a completely arbitrary outcome of the fact that someone just looked at it. It doesn't really happen for a reason, it just happens. Consequently, even though both systems and both experiments are identical, their wave functions may collapse to two different values once they are observed by their respective observers.

It started out as a sort of half-joke of Schrödinger's, a mocking finger pointing out something "ridiculous," but the scientific world was compelled to take it seriously. It has become a central feature of quantum mechanics, its character fading from ridiculous to intriguing, and is discussed today with that intellectual coolness reserved for notions that are fairly obvious. It is perhaps the reflection of a vague resentment that we have given the whole issue the rather prosaic title of "the measurement problem."

This "problem" is scandalous for two reasons. Firstly, it makes quantum theory seem not only strange but suspicious. It says that the world really is probabilistic and uncertain and indeterminate, but only when no one is looking. Moreover, because this collapse of the wave function manifests itself as something completely arbitrary, it is essentially unexplainable. Again: it doesn't really happen for a reason, it just happens. If we can characterize the observer, and we can characterize the physical system, it seems that we should be able to characterize the relationship between the two. And yet a clear understanding of this relationship, and any explanation of why there should be this collapse in the first place, is conspicuously missing from quantum theory.

True, quantum theory "explains" the result of a particular experiment in terms of the probability of its occurrence, as calculated using Schrödinger's equation. But relying on this sort of explanation has strange consequences, even for the concept of probability itself. From the classical point of view, probability is a way of handling complex processes on a macroscopic scale. It uses statistical relations to calculate the expected outcomes of a number of sample measurements, and a minimum number of samples is required for a calculation to be considered "statistically valid." For example, I can say that when I flip a coin it has a 50 percent probability of being heads (or, more correctly, a probability of 0.5). By this I mean that, if I flip the coin many times, it will turn up heads about half of the time. Or similarly, if I flip many coins at once, about half of them will turn up heads. What actually causes each coin to turn up either heads or tails is a combination of various factors—the initial force on the coin, its angle relative to the person's hand, air resistance, and so on. In principle, with enough computing power one could bypass probability theory altogether and determine which coins will turn up heads by directly calculating these factors on each coin.

This is not the case for a "quantum mechanical" system such as, for example, an electron orbiting the nucleus of an atom. It turns out that an electron only exists in certain discrete orbits, or energy states, within an atom. Quantum mechanics can be used to calculate the probability of the electron being in one of these states. But remember that the wave function is a "complete" description of the state of a system. Unlike classical probability, we are not referring to a statistical sample of many electrons but to the energy state of that one electron only. The particular orbit of the electron is not determined by external forces or factors, but by its inherent probabilistic nature. Classical physics would like to think of the electron orbiting the nucleus as it does a satellite orbiting the earth. The orbit of a satellite is a function of its angular momentum; a faster velocity will result in a wider orbit. By contrast, there is no external force that would allow the electron to be in some intermediate energy state. It only exists in certain states, and these can *only* be calculated as probabilities. When we are asked what actually causes the electron to be in a particular state, we are at a loss. Because of this Bohr insisted, much to Einstein's dismay, that quantum theory forces us to question even the fundamental concept of causality.

There is therefore a paradoxical relation in quantum theory between the complexity of a system and its probabilistic nature. As one moves away from a macroscopic system of complex components toward a quantum mechanical system of elementary particles, the system strangely becomes *more* probabilistic. It is as if probability were attempting to acquire an independent existence by denying its own meaning.

The second reason that the measurement problem is so scandalous arises from the natural tendency to equate the notion of an observer with that of a human being. After all, an observer is the subject of an observation (as opposed to the object of observation), and a subject is subjective, which implies a human person. But the implication that the state of a physical system should be so intimately dependent on the person observing it, should in fact collapse as if under the weight of a glance, offends our scientific sensibilities, which have always imagined that the world is majestically indifferent to our humanity. We have grudgingly accepted the idea that the physical world depends on an observer, but the notion that it depends on a human being is going too far.

Nonetheless, the fact is that it was the introduction of an inherently nonprobabilistic conscious mind into a probabilistic physical system that created a measurement problem in the first place. It was not the Geiger counter in Schrödinger's experiment but the person looking at the Geiger counter (or at the cat) that made the difference. The central role of human consciousness has been emphasized by a number of physicists over the years, including John von Neumann, Eugene Wigner, John Wheeler, and others. Wigner created one of the more provocative images by introducing a subtle variation to Schrödinger's thought experiment, in which the cat is replaced by a human being. These arguments generally take the implications of this experiment to the extreme, suggesting that the physical world itself is not really real, but is in a sort of undefined state of superposition unless and until a human person observes it and brings it into concrete reality.

Of course, this has opened the door to a wide range of metaphysical arguments, some of which admittedly push the limits of credulity. The implications of "quantum consciousness" seem to suggest that subjective human beings somehow *create* reality itself, and in fact Wheeler has coined the term "participatory anthropic principle" to express this basic idea. In other words, not only does the tree make no sound when it falls in the forest, it does not even exist unless someone is there to observe it. The majority of the scientific community is generally put off by such dubious metaphysical implications. But this does not make the measurement problem any less of a problem.

The term "consciousness" is in fact misleading. It seems to me that the key characteristic of the observer in the measurement problem is that he or she is *self*-conscious or self-aware, as suggested, for example, by the physicist Henry Stapp.[4] Only a self-aware observer is capable of observing *himself* as well as the physical system.[5] A person is therefore essentially in a perpetually collapsed state, so to speak, which is ultimately a nonprobabilistic state.[6] His ability to observe himself and the physical system, and himself *within* the physical system, means that the system must be related to him in some nonprobabilistic way. For quantum mechanics this means that the system suddenly stops being probabilistic—it collapses to a single well-defined state—as soon as the person observes it.

The general trend in physics, however, is to avoid the subjectivity of the subject, and this has degraded over the years into a purely materialistic understanding of the observer. Most quantum theorists deny any technical difference between a human observer and any inanimate measurement device such as a clock or a ruler. It seems to me that this attempts to deal with the measurement problem by avoiding it altogether. The measurement problem, after all, is ultimately a reflection of the old problem of relating epistemology to ontology—the knowledge of the observer to the state of the object. It is no coincidence that most of the greatest figures in quantum theory have felt compelled to write books and papers on the philosophical implications of their work. But, of course, one does not speak of the subjective knowledge of a clock or the epistemology of a ruler, which gives the impression that we can simply talk about clocks and rulers and ignore everything else. Even if things like subjectivity and knowledge are dismissed as philosophical or unscientific, they nonetheless pertain to the specifically human observer. In this case, at least, these concepts may be extremely relevant to understanding the measurement problem.

The decision to either accept or reject the uniqueness of the human observer is mostly a matter of opinion. There is no solid mathematical or "scientific" proof in quantum theory for either position. Most physicists that reject the observer's humanity admit quite freely that they do so basically from a personal bias, or because they don't see a reason to complicate things with all this business of minds and consciousness.[7] It is interesting that the scientific community feels comfortable with such blatantly subjective arguments but insists on more objectivity from anyone defending the opposite opinion. It seems intuitively obvious to most people that there is something unique about the human person that categorically separates him from inanimate objects, and this is evidenced by the fact that scientists keep arguing about it. If anything, one would think the onus is on the materialist to prove otherwise.

As we've said, there is no conclusive argument one way or the other. But if we insist on reducing the human observer to a purely "objective" measurement device, we run into an insurmountable obstacle. After drawing every physical and mathematical parallel between human and inanimate observers, there is in the end one difference that remains: the human observer can be wrong. This dubious distinction is absolutely unique to the subjective human being and is usually the reason given for ignoring him in an experiment. Subjectivity is a source of error, but precisely because of this it is also the source of the human observer's uniqueness.

This ability to be wrong is not true of any nonconscious measurement device. We are used to speaking about a certain meter or gauge being wrong, but by this we mean that the device did not operate according to a (human) person's expectations or intended design. Perhaps it was installed incorrectly in the system or was calibrated for a different experiment. Either way the error is not in the device itself; it is in the experimenter's knowledge of the device. In fact, we know we have found the source of the "error" when we discover a situation for which the device's measurement is *correct*: if the calibration is off, then a correct measurement by the device would produce the "wrong" result for this experiment. A device only measures what it is built and set up to measure, say, the number of photons striking a screen or the presence of a magnetic field. But it does not know anything about the object producing these effects, and consequently can say nothing right or wrong about it. As the philosopher Martin Heidegger puts it, a purely objective or "mathematical" observation attempts to establish the being of a thing by "skipping over" the thing itself and recording only the facts of the thing.[8]

For example, a Geiger counter measures the energy generated by a nuclear particle striking a detector, and this is used by nuclear engineers to determine the presence of radioactive material such as uranium-235. But if we could ask a Geiger counter, "Is there a piece of uranium-235 in front of you?" it would have no idea what we were talking about. It would never ask, "Is there really an object in front of me, and if so, what is it?" This is the nature of any purely objective measurement device. It cannot give a wrong answer because it never asks any questions, it only makes statements. Only a human being asks questions.

One may object that we are merely referring to the fact that human beings *interpret* their observations and may make mistakes when doing so. As strictly objective observers, however, they are no different from any other measurement device. The implication is that one can separate a person's subjectivity from the act of that person's observation. But this is a bit like saying that elephants are no different from mice, once you remove the fact that elephants are larger, have trunks, and don't have whiskers. If we remove subjectivity from the human observer, we, of course, have no problem at all. We also no longer have a human observer.

The measurement problem demonstrated by Schrödinger's thought experiment is not the result of the person's interpretation after having looked inside the box. The problem arises because the human observer, unlike the Geiger counter in the box, observes his own observing, and therefore *knows* what he is looking at. This "knowing" is a judgment on the truth of the observation, a ruling on the question of whether the cat is *truly* alive or *truly* dead. It is not a separate operation that is done after the fact, but a fundamental characteristic of human observation itself.

I will speak more about knowledge and truth later on. For now I simply want to preserve a certain clarity of intuition—that intuition that allows us to see, even in the face of very clever arguments, that a cat is still a cat. Discussions and mathematical representations of the measurement problem have gotten more and more complex over the years in efforts to understand it. But in general we are not bothered so much by the complexities of quantum mechanics, and in fact take some comfort in them, in the consoling notion that complex problems should indeed be difficult to understand. No, what is especially striking about the measurement problem is not its complexity but its stark simplicity, and this is what really bothers us—that something as straightforward and innocuous as looking at an object should be the cause of such strange effects. It's a bit silly and unbelievable, like a cheap magic trick. It's true that a certain natural phenomenon may seem like magic, an object of fear and wonder, if the underlying complexities of the physical laws are not known. A magnet appears to cause a piece of metal to move "for no reason." The collapse of the wave function has a similarly simplistic and magical quality, with electrons occupying one state or another "for no reason." If we think of the act of observation or measurement as tying a rope to the object and holding fast to the other end, the mechanism of collapse shows that it is a magician's trick rope that somehow becomes discontinuous, or even suddenly tied to something else, when our attention is distracted for a moment. We have always relied on the continuity of this rope, however long and twisted philosophy might make it, to tie us directly to the object of our attention.

This apparent simplicity derives from a basic assumption that is taken for granted in both classical and modern physics. Science has always assumed a direct linear connection between the observer and the object of observation. I do not mean linear in the strict mathematical sense, as in a linear measurement operator in quantum mechanics. I mean that there is conceptually a kind of line tracing out the observer's direct relationship to the object of observation. Any physical object is considered as having a certain inherent "observability" or "measurability," and the observer measures the object accordingly. In other words, the act of observation is primarily and fundamentally defined by the object of observation and the observer observing it. This almost tautological assumption leaves little room for mystery, and it has been the bane of modern physicists to find *the* mystery precisely *there*.

It may seem strange to claim that science has a simple understanding of observation when so much effort over the last century has gone into complex analyses and descriptions of this central concept of modern physics. When one surveys the landscape of scientific theories such as Kuhn's "paradigms" or Bruno Latour's "hybrids," or the various interpretations of quantum theory such as Bohr's "complementarity" or Hugh Everett's "many worlds," the word "simple" does not come to mind.[9] But we should not confuse the theories themselves with the thing they are trying to describe. The term "simple" here does not refer to the various twentieth-century theories that have tried to explain things, but to the underlying subject-object relationship that tends to make them all seem overcomplicated—what could be called a "linear" or "object-oriented" models of observation in the sense described above. This awkward tension between the simple structure of observation and the complicated theories of observation is precisely what lies at the heart of the whole discussion.

The distinction becomes clearer when we remember the insistence of modern scientists that an observer is merely a measurement device, whether alive or inanimate. In the case of an inanimate apparatus, there is no question of anything other than a simple and direct (i.e., "linear" or "object-oriented") connection between the observing apparatus and the observed object. There are no group dynamics here, no communal perspectives or imposed paradigms. An inanimate measurement apparatus does not have a theory of observation. It interacts directly and unambiguously with another object by physically contacting it, through some physical medium. If there is no significant difference between this apparatus and a human being, then it means there is nothing fundamentally different about this direct connection to the object of observation. We may be forced to explain other related epistemological or perspectival factors that our humanity brings into the situation, but all that is ultimately in contrast to the simple "linear" connection to the object.

In classical physics, nature is independent of man, and the connection between the observer and the observed is incidental and independent of the workings of natural laws. Because of this independence from nature, the observation of any one observer is necessarily independent from that of any other observer, and this in turn implies the independence between the observers themselves. Even if classical physics is not always precisely correct, it is at least logically consistent in its understanding of the world and man's place in it.

Modern physics, however, claims a crucial relationship between the observer and the observed. Far from being an independent or incidental side-feature of an experiment, it is an integral part of the physical process itself. And yet the lines connecting the observers to their respective objects of observation remain completely independent, and one observer in principle has nothing directly to do with another. Einstein called this the principle of relativity, and it is a fundamental postulate of his famous theory. In quantum mechanics, this independence between observer-observed interactions is what allows the two identical experiments mentioned earlier to collapse to different states. If the two wave functions were not independent of each other, they would by definition be part of the same system and form a composite probability distribution, which would then collapse to a single value.

So despite the observer's entanglement with the world around him, he remains somehow completely free from the entanglements of others. He is part of his own experiment, but somehow not part of anyone else's experiment, at least not as an observer. Modern physics, therefore, does not really assume anything fundamentally different about the linear or object-oriented nature of this connection between the subject, as observer, and the object. The most important philosophical contribution of quantum theory, I believe, is in showing just how problematic this simple assumption really is.

In fact, in recent years this line connecting the observer to the object of observation has been strained almost to the breaking point. The first big tug was made in 1935 by Albert Einstein, Boris Podolsky, and Nathan Rosen in a paper entitled "Can Quantum-Mechanical Description of Physical Reality Be Considered Complete?" In this paper they presented a thought experiment involving two objects, or two components of a system, that interact in such a way as to necessarily have opposite states. That is, if object 1 (call it O1) is in state A, then object 2 (call it O2) must be in state B, and vice versa.[10] No measurement is made on the system, so there is no knowledge of which object is in which state. Now let these two objects go off in different directions and travel away from each other. After a very long time, perhaps many centuries, we can justly say that O_1 and O_2 are no longer related to each other in any way and should be considered separate and independent systems. At this point, let's allow an observer at O_1's location to measure the state of O_1, and let's assume it turns out to be in state A. If this is the case, then we know with 100 percent certainty that when O_2 is measured it will be in state B. But we know this even before any measurement is made on O_2. Quantum theory says that we cannot know the state of O_2 (or any object) with 100 percent certainty unless and until we measure it. How then is this possible?

Einstein, Podolsky, and Rosen concluded that there must be some other factors acting on O_1 and O_2, some preexisting ontological property that quantum theory does not take into account. It is this "hidden variable" that really determines the states of O_1 and O_2, regardless of whether or not they are observed. The alternative is that there is some sort of unknown connection between O1 and O2 that carries information from one to the other, effectively clueing O2 into what just happened to O1 so it knows to what state it should collapse. But this information would have to travel instantaneously over an arbitrarily large distance, faster than the speed of light, which would violate a fundamental rule of special relativity. Einstein dismissed this possibility offhand, calling it "spooky action at a distance." The paper was therefore considered a solid defense of "hidden variables theory," and for several decades this was the most viable alternative to the Copenhagen interpretation of quantum theory.

In 1965, however, things took a turn for the strange again. The physicist J. S. Bell worked out a set of equations (known as the Bell inequalities) that provided a way to effectively detect the presence of hidden variables. In principle, an experiment could be set up in such a way that the results would either confirm or disprove hidden variables theory, depending on whether or not they satisfied the Bell inequalities. It took a couple more decades to actually design and perform these experiments in the laboratory, but when they did, the results were irrefutable. They showed conclusively that there were in fact no such things as hidden variables.

This means that Einstein's dreaded alternative must actually be true. There must indeed be some "spooky" connection between different experiments, some sort of underlying physical or metaphysical space in which physical objects are entangled. This is known as the EPR effect (referring to the authors of the original 1935 paper). It is another one of those grudgingly accepted yet unexplained features of quantum theory that many physicists find even more disturbing than the collapse of the wave function. It means that the universe is "nonlocal," which is even further away from classical physics than most quantum theorists were willing to go. The concept of locality, which had been taken for granted until now, assumes that two physical systems that are separate from each other are independent of each other. The EPR effect forces us to abandon this intuitively obvious notion in favor of a universe with mysterious strings being pulled behind an invisible curtain.

But note that while the EPR effect admits to a connection between two different experiments, it sees no connection between different *observers*. Observer 1 is associated with object O1 and observer 2 is associated with object O2. We speak of the state of O2 depending on the state of O1, we do not speak of O2 or observer 2 depending on observer 1. The problem exists already even before there is an observer 2, in the fact that the state of O2 is determined without any observation at all. In other words, even in this surprisingly interconnected universe, we still manage to maintain the independent "linear" connection between the observer and the object of observation. And this is in fact why the EPR effect is so disturbing. If the observers were somehow related, were images of each other or had some inherent ontological relationship, this would make the connection between their observations more palatable, if not intuitively obvious. But this is not considered in the way the EPR thought experiment is formulated. We are therefore forced to imagine some sort of sympathetic relationship between the objects themselves that has no explanation either in relativity theory or in quantum theory.

So over the years quantum theory has proven to be the most frustrating kind of opponent—the kind that accepts every attack and insult with a defiant smile. Schrödinger calls quantum theory ridiculous, and quantum theory responds by saying, "Yes, I am indeed just that ridiculous." Einstein calls it spooky, and quantum theory says, "Quite right, I am indeed spooky like that." This does not strike us as the sober attitude of a rational scientist, or even the imperturbable logic of a Socrates. It is more like a stubborn child who knows he is doing something wrong but refuses to admit it's wrong. "So I'm ridiculous. So what?"

We have been discussing these ridiculous and spooky issues—the arbitrary collapse of the wave function, the paradox of probability, the interdependence of independent observations, and the confusion over what an observer is in the first place—as problems of quantum theory. But it is more correct to say that these are symptoms of a more fundamental problem. This becomes painfully obvious when we consider the knee-jerk reaction of some modern physicists to the annoying metaphysical questions that these issues keep dredging up. The mathematics of quantum mechanics works very well, regardless of their specific interpretation, prompting the now popular motto, "Shut up and calculate!"

This is meant, of course, to express a kind of pragmatic attitude toward the practice of science. But is there any truly scientifically minded person who is satisfied with such an attitude? I don't think so. If our current understanding of the world produces metaphysically dubious results, and our only response to them is stubborn avoidance, then the rational (and scientific) thing to do is take a step back and look again. Clearly, we're missing something.

2
Modern Observation through Medieval Eyes

As a theory of physics rather than metaphysics, remaining true to the notion of scientific objectivity, quantum theory has felt it necessary to avoid the subjectivity of the observer. We are expected to take the subject seriously, but not the subject's subjectivity. With a simple linear model of observation this leaves only one option—turning away from the subject leads one straight to the object. If one tries to avoid metaphysics in this way, subjectivity is effectively transferred to the physical system, as a kind of property of nature, or the "observability" of nature. But while subjectivity in a human person poses no logical or philosophical dilemma, subjectivity in nature, in the absence of any conscious will, is reduced to probabilities and randomness. It is this strange image of nature against which Einstein and many of his contemporaries rebelled so strongly, and it is also, ironically, what has led to some of the strangest metaphysical ideas.

The philosopher Thomas Nagel has explored in great detail the relationship between the subjective and the objective, and his philosophy is largely built on the insistence that human subjectivity is unavoidable in any serious model of reality. "The subjectivity of consciousness is an irreducible feature of reality—without which we couldn't do physics or anything else—and it must occupy as fundamental a place in any credible world view as matter, energy, space, time, and numbers."[1]

Nagel does not delve into quantum theory, nor has he made many comments on its philosophical implications. But in a discussion on human freedom, he provides the following description of "our ordinary conception of autonomy," which sounds strikingly like a description of the collapse of the wave function.

> Although many of the external and internal
> conditions of choice are inevitably fixed by the
> world and not under my control, some range of

open possibilities is generally presented to me
on an occasion of action—and when by acting
I make one of those possibilities actual, the final
explanation of this . . . is given by the intentional
explanation of my action, which is comprehensible
only through my point of view. My reason for
doing it is the *whole* reason why it happened,
and no further explanation is either necessary or
possible. (My doing it for no particular reason is a
limiting case of this kind of explanation.)[2]

A range of possibilities that, when acted on by the subject, is reduced to one actuality "for no particular reason." Is this not a perfect description of the arbitrary and noncausal collapse of the wave function by an observer? It is, but with one important difference. The "final explanation" for the collapse is given, not in terms of the probabilistic dynamics of the object, but the intentional action of the subject. These strange behaviors of nature, the fickle wave function and the irreverent disregard for causes and effects, do indeed seem "ridiculous" as features of a well-ordered universe. They shocked and confused the scientific community with an image of the world that seemed intriguing to some and unbelievable to others. But this very same image is not shocking at all, or even particularly modern, as a description of free, human, intentional action.

In fact, there have been several studies in recent years using the mathematical logic of quantum mechanics to model cognitive and decision-making processes in the human mind.[3] Classical models of cognition have been shown to be incompatible with various "empirical puzzles" observed in human behavior. This is driving a growing number of psychologists to turn to quantum mechanics to better describe these strange features of human thought. Central to these analyses is the concept of collapse from a probabilistic distribution of options to a single realized decision.

The goal of these efforts is to show how the human mind acts like a quantum mechanical object. But this of course means that a quantum mechanical object acts like a human mind. More generally, the relationship between physical object and human mind has features that today we would call quantum mechanical. This may sound like a very modern idea, but it has actually been known for a long time. Centuries before either Nagel or Bohr discussed these issues, before the modern confusion over the meaning of "observer" and the materialism of post-Enlightenment scientism, many of the unique features of quantum theory were already present in the Scholastic philosophy of Thomas Aquinas. This was first noticed by Werner Heisenberg (actually in terms of Aristotelian physics) and has since been analyzed in different ways by a number of philosophers and scientists. While these analyses vary to different degrees in their approach, the key concept underlying all of them is that of *potentia*.

According to the Angelic Doctor, what we are calling observation in this study is not merely a passive peripheral phenomenon, as classical physics would have it. There is an active and a passive element involved. More specifically, Aquinas maintained that all objects in the world are passively ordered toward a knowing mind—that is, all objects have an inherent knowability, much like the measurability assumed by the physical sciences, which gives them the *potential* to be known. This potential knowledge is actively converted to actual knowledge by a human mind in the act of observation.

In fact, we have here not merely a divergence from classical physics but an essential reversal of its most basic assumption regarding the "passiveness" of the observer. For Aquinas, it is the physical object that is passive and the human mind that plays the active role in any observation. The philosopher Josef Pieper describes Aquinas's thought as follows:

> The objects in and of themselves do not actively establish a relationship with the mind; rather, they are passively "ordered" toward it. It is the mind that in its most specific activity "relates" to reality; more precisely, it is the mind that changes an already existing but only potential relationship between objective reality and subjective cognition into actual fact.[4]

This action of the mind is actually the acquisition of the "essence" of the object, an idea that has its basis in the Aristotelian concepts of form and matter. Any object exists at all by virtue of its matter, but it exists as itself (as opposed to anything else) by virtue of its form. Matter determines *that* it is—the primary "being" of the object—while the form determines *what* it is—the "essence" of the object. The essence of a particular statue, for example, is not the material itself, which existed independently of the statue even before the statue was made. It is the form given to it by the artist that makes it a statue and gives it its "statueness"—its essence as a statue.

In fact this form is really comprised of two types: substantial and accidental. Substantial form is what is more properly associated with the essence of a thing—the statueness of a statue. But statues can have different shapes and sizes and still be statues. This aspect of a thing that can change or be different without changing the essence of that thing is its accidental form. But at any point in time an object must have both substantial and accidental form, along with its matter.

What makes the human intellect unique is that it has the ability to take on or assimilate the form of the thing that it observes. Aquinas says: "Beings endowed with the ability to know are distinct from those not so endowed inasmuch as the latter do not have any other form but their own, while the former are capable of having also the form of the other being."[5] The act of observation, in which potential knowledge of an object is converted to actual knowledge, is precisely the assimilation of the form of the object by the knowing mind. Moreover, the form is not merely an image or representation of the object in some vague sense, but is precisely the same form, and therefore *the* form, of the object. What exists in the mind is not the object itself, of course, because it does not contain the matter of the object. But the actual form of the object, the thing that makes it that particular object, is indeed contained in the mind. For Aristotle (and Aquinas), there is an essential equivalence between the object of knowledge and the knowledge of the object.

The notion that nature exists both in potentia and in act is very important to Aquinas, and we find its analogue in the physical sciences as well. Both states are equally real, in the same way that both potential energy and kinetic energy comprise the total real energy of a system in physics. Kinetic energy is the energy of a system actualized in motion, and potential energy is the energy stored in a system that gives it the potential for motion. For example, a compressed spring has real potential energy stored in it, energy that requires a specific act of force to restrain it (i.e., keep it compressed). If that force is removed, the energy is released as kinetic energy through the expansion of the spring.

All of this is central already to classical physics, and as such is understood independently of any observer. Nonetheless, we can speak in classical terms of the observability of these forms of energy. For example, we can say that kinetic energy is energy in act—that is, in the act of manifesting itself as motion. Without knowing anything specific about an object itself (e.g., its position or velocity or mass), one can see directly that it has energy by virtue of the fact that it is moving. One can observe motion and recognize it as the manifestation of kinetic energy. In other words, kinetic energy is inherently observable.

By contrast, potential energy is energy that is *not* in act, and is therefore inherently not observable. Without knowing anything specific about an object (e.g., whether the spring is compressed or merely a short spring), one cannot tell whether the object has any energy at all. An object does not show its potential unless it does so indirectly by converting it into something actual—something that can actually be observed.

Of course, to a classical physicist these distinctions are purely academic. The presence or absence of an observer, and whether or not a particular physical property is observable in principle, does not affect the physics itself or its effect on the object. This changes, however, with relativity theory, in which the implications of observability take an important step forward and an important step back toward Thomistic ontology.

By considering the effects of a moving observer on the laws of physics, Einstein came to what is now the most famous conclusion in modern science—the effective equivalence of energy and mass, or $E = mc^2$. The constant c, the speed of light, is just a number, specifically 2.998×10^8 m/s. The equation states, then, that energy and mass directly imply each other, and are in fact interchangeable. For example, the process of nuclear fusion powers and sustains our sun by converting nuclear mass directly into energy, and physicists at CERN convert the kinetic energy of accelerated particles into entirely new particles of mass that weren't there before.

Because of this equivalence we can extend the notion of the observability of energy to the observability of mass. That is, if total energy is comprised of observable (kinetic) and unobservable (potential) energy, then one can imagine there must be corresponding components of observable and unobservable mass. We are not speaking of mass (i.e., an object) that just happens to be unobserved at the moment, but an object that is in principle unobservable while in its *potentially observable* state. If an object in the world is being observed, the implication is that its "potential mass" (i.e., its potential observability) has been released (i.e., actualized), in the same way that kinetic energy is the release of potential energy.

Now, mass is basically the amount of matter an object has, independent of the specific material, size, or shape of the object itself. We may therefore think of energy in Aristotelian terms as a sort of "prima materia," or pure matter—that is, matter without form. The conversion of energy into mass is then really nothing more than the attribution of a specific form to a certain amount of energy/matter, resulting in a localized object with mass.

The opposite condition of reality—form without matter—is precisely what is found in the knowing intellect. If an object comprises matter and form, then the existence of that object implies the coming together of energy and intellect. In this case, the physical world in the absence of an intellect must be an unobservable world of formless energy. It can take on a specific measurable form only by coming into contact with an observing intellect.[6] This sounds very much like a restatement of the measurement problem, except that there is nothing very problematic about it. There is no strange or arbitrary collapse of reality, only the natural coming together of matter and form to produce an observable physical object.

Similarly, the potential knowability or "potency" of Thomistic ontology transforms into actual knowledge only in the presence of a human observer. Pieper argues: "It pertains to the nature of all things that by being known they are being 'grasped' and thus become the prey and property of the knowing mind. The nature of the mind, on the other hand, lies in its ability to 'grasp' in turn all things."[7] Moreover, Aquinas agrees with modern science that, while an observer is required for actualization, an object's potency exists already in nature, outside the mind of the observer. This knowability of Thomistic ontology exists always as a potential relationship with the observer. When the object comes in contact with an observer, its preexisting potential is actualized, and its potential observability becomes something actually observed. This precludes any notion of physical reality itself being merely a figment of the observer's imagination, the product of some "quantum consciousness": "A mutual correlation between *being* and *mind* exists already prior to any actual perception. The object-subject relationship, in traditional ontology, is not at all 'a creation on the part of the mind'; it precedes any activity of the mind."[8]

It seems, then, that we can reach very "quantum mechanical" conclusions without invoking quantum theory at all. All we need is Aristotle, Aquinas, and Einstein. But, alas, quantum theory is not so easily tamed, and upon closer inspection we find its usual resistance to easy interpretations. First, we must recall that the form of the object in its unobserved state of *potentia* is a probability distribution, a well-defined range of physical states, each of which is no less real or actualizable than the other. Therefore, contrary to what we concluded only a moment ago, the object in potency is not unobservable because it has no form. It is unobservable because it has *too many* forms, and therefore no single form to observe.

More importantly, for Aquinas, what exists in nature "prior to any actual perception" is not a probabilistic state of the object itself but its potential relationship to the subject. The object in relation to the subject, the "correlation between being and mind," is precisely *not* probabilistic in the quantum mechanical sense—it involves the knowledge of a single state of reality. The potential reality of an object is not its potential to be observed in one state or another, it is the potential for the one state of the object to be observed at all (that is, to be known by a human intellect).

The act of observation is the correlation of the intellect with the essence of the object. The essence of an object is not what the object could be or might be but only what it is, and this is what correlates (is in fact identical) to the form in the intellect. All this implies that the form is only one thing, one reality corresponding to any particular object.

Any other form of that object is not its true form, and therefore not real.

If the object in potentia does have a range of forms, as quantum theory suggests, how do we know which form will correlate to the form of the knowing intellect at the moment of observation? We don't, as Heisenberg made mathematically clear. By contrast, Aquinas tells us that all of nature is ordered toward a knowing mind. Everything that is real, that exists in the first place, is either actually known or potentially known. The notion that nature can comprise realities that suddenly cease to be knowable once another reality is observed goes directly against this ontological principle. While quantum theory focuses on the inherent uncertainty or "unknowability" of nature, Aquinas focuses on its inherent knowability.

It seems to me that this aspect of Thomistic thought has not been fully appreciated by those who have focused on the parallels between Aquinas's *potentia* and the quantum wave function. The philosopher Roger Paul, for example, states unequivocally, "Therefore, the potentialities that quantum mechanics define in the wave function are knowable because through measurement they become actualities."[9] But the problem is that they do not *all* become actualities, only one of them does. What about the rest?

In fairness, Paul does associate these potentialities more specifically with the accidental forms of the object. His thinking is that the accidental form is independent of the essence and the matter of the object, and so does not directly correspond to the thing itself. This would seem to avoid the ontological dilemma, since an object can indeed have different accidental forms and still be itself. But this does not really solve the problem because, just as an object can have only one substantial form, it can have only one accidental form at a time. An electron can be here or it can be there and still be an electron. But if at any point in time there is an electron here and an electron there, we are talking about two different electrons. This basic ontological principle cannot be reconciled with the wave function of a single electron and its claim to scientific completeness.

What started out, then, as a happy agreement between Scholastic philosophy and modern physics has ultimately pointed to a fundamental disagreement. But both the agreement and the disagreement are significant. The notion that unobserved physical reality is essentially a potential reality plays a key role in both models of the world. In Thomistic philosophy it plays the role of a hero, whereas in quantum theory it plays the role of a villain. For Aquinas, the potential reality of nature is a continually opening door, leading to an ever deeper relationship between the human intellect and the true essence of reality. In other words, Aquinas draws a direct connection between the epistemology of the observer and the ontology of the observed. Quantum theory shies away from any such connections and points to epistemological limitations while generally waving its hands at ontological questions. It is more like the dropping of a veil than the opening of a door.

How, then, are we to reconcile the Thomistic understanding of knowability and truth with the multiplicity of forms that seems to make up the essence of physical nature? Ironically, the problem is not in where the Thomistic and quantum models disagree, but precisely in where they agree. More specifically, both assume that the act of observation involves a direct linear relationship between the observer and the observed, in the sense discussed earlier. This of course poses no problem for Aquinas and serves to strengthen the ontological connection between the human intellect and the truth of the created world around it. So close, in fact, is the connection between observer and observed that they literally share the very essence of the object itself. For quantum theory, however, this linear model of observation has been the bane of its philosophical development. We have seen how it has uncovered a host of metaphysical issues under this deceptively simple assumption, and it is doing the same now with Scholastic ontology. As long as we hang on to this linear model, the multiplicity suggested by quantum theory is incompatible with the philosophical realism of Aquinas.

The obvious thing, then, is to do away with this linear structure of observation and replace it with something else. But what? How do we simply replace an assumption that has been taken for granted in science and philosophy since the earliest stages of both?

In fact, we need not guess blindly. The answer will already be fairly obvious to students of the anthropological school of thought known as mimetic theory. These scholars will recognize immediately the same basic problem here that is addressed in mimetic theory—the problem of a direct, independent linear relationship between subject and object leading to dubious conclusions that do not agree with what we actually observe in the real world.

Of course, the subject in mimetic theory is a human being, and this may seem like a serious limitation considering the debate over this topic. If the quantum mechanical observer is really nothing more than a measurement device, and there is no significant difference between a person and a ruler, then the comparison between quantum theory, Scholastic theory, and mimetic theory is a purely metaphorical exercise. But it seems strange to insist that people are rulers when they have proven to be particularly bad at following rules (such as the rules of causality and repeatability). Why impose this arbitrary constraint, which goes against all experience and intuition? For the sake of scientific objectivity? Is it not ironic that we should stubbornly defend a particular subjective bias in the name of objectivity? This attitude has done little so far to clear up the symptoms described in the first chapter, so it may be worthwhile to see where the opposite assumption leads. If it leads to a model that is less philosophically and metaphysically dubious than our current understanding, then the truly unbiased and scientific thing to do is to take that assumption seriously.

As a minimum, we must acknowledge the basic fact that the observer *can* be a human being, and often is. In this case, even if it's a special case, we are not dealing with a mere metaphor or analogy. In this case, the subject in the subject-object relationship of quantum theory is the same as that of mimetic theory. And as the first chapter points out, this subject is unique for at least two reasons: first, it does not lend itself to the standard linear model of observation, and, second, it can be wrong. Precisely these two features capture the attention of the mimetic theorist.

The central concern of mimetic theory is not the nature of knowledge or measurement per se but the nature of desire. It is more interested in analyzing origins than final states. And it explicitly acknowledges what most quantum theorists stubbornly deny: the uniquely significant role of the human subject. But such differences are primarily differences in context, in the space within which we investigate and work out the details of the problem. The problem itself has yet to be fully understood, and we will see that mimetic theory helps us do precisely that.

3
The Interdividual Observer

Mimetic theory is primarily a theory of human desire. At first glance this may seem a rather restricted scope of study, especially if the word "desire" is taken to mean merely sexual or appetitive desire, and these in turn as merely one or two components of a repertoire of psychological emotions. But desire here should be understood in the broader sense as that which motivates attitudes and actions in general. My desire is not merely a statement of what I want but *why* I want it and under what conditions that wanting is created and sustained. A cheap romance novel may present desire as a sort of static magnetic attraction toward some object of lust or greed, but that is what makes it cheap. More intelligent novels trace out the subtle and complex elements within a network of relationships that define the dynamics of desire.

Behind every human relationship, every act of charity or cruelty, every cultural institution and tradition, there lies this dynamic of human desire. Purely appetitive desires—for example, I want food because I am hungry—can indeed be understood more statically as an attraction between the desirer and the object of desire, and this is what generally motivates the attitudes and actions of animals. But it is clear to the good novelist that human desire goes well beyond this.

Unfortunately, this has not been so clear to many who have taken it upon themselves to characterize human relationships. The nature of human desire has typically been understood in a "linear" way, as a direct linear connection between the desiring subject and the object of desire. The object has some inherent "desirability"—some property which makes it desirable—and consequently the subject desires it. For example, we desire a new car because it is more luxurious than the car we currently have, and this property makes it more desirable. We desire a particular person because he or she is attractive, and therefore "attracts" our desire. In general, people act a certain way because they desire certain things (food, security, power, etc.), and they must desire these things because of certain inherent properties that offer some sort of benefit to the person. It would be irrational to desire something that offered no benefit or was even detrimental. Such things would have no reason to be desirable.

And yet history is filled with examples of human behavior violating this simple principle. The complex web of human relationships has, of course, been the primary object of observation for novelists, poets, and playwrights, and those particularly insightful observers have shown how contradictory and even self-destructive human desire can be. Moreover, the great contribution of these artists has been to show that this "irrational" behavior is not merely confined to psychotic or psychologically disturbed people, but applies to some degree to all of us as human beings.

Dostoyevsky's "underground man" serves as a classic example of the sort of behavior that drives people to throw obstacles in their own way:

> I am a sick man. . . . I am an angry man. I am an unattractive man. I think there is something wrong with my liver . . . I am not having any treatment

for it, and never have had, although I have a great respect for medicine and for doctors. . . . No, I refuse treatment out of spite. That is something you will probably not understand. Well, I understand it. I can't of course explain who my spite is directed against in this matter; I know perfectly well that I can't "score off" the doctors in any way by not consulting them; I know better than anybody that I am harming nobody but myself. All the same, if I don't have treatment, it is out of spite. Is my liver out of order?—let it get worse![1]

Similarly, Camus's main character in *The Stranger* destroys his own future by murdering someone he does not even know with a gunshot, adding four more almost out of indifference, "like knocking four quick times on the door of unhappiness."[2] Don Quixote is, among other things, a brilliant caricature of the "insanity" that can characterize such human conditions as bravery, honor, and jealousy. It is an insanity that can reach surprisingly dangerous proportions, as evidenced by the paranoid narcissism of Stalin or Hitler.

A detailed study of human behavior and relationships is obviously beyond the scope of this monograph. Suffice it to say that the straightforward understanding of human desire as a linear pragmatic relation between subject and object has proven to be simplistic, and this has not gone unnoticed. It was noticed, for example, by Sigmund Freud, who understood that a consciously pragmatic model of desire could not account for the types of behavior often observed in people. Hence the introduction of the *un*conscious mind. It is irrational to consciously desire something that will do us harm, but a latent mechanism hidden below the cognitive radar of conscious thought, and rooted in causes that occurred perhaps years ago, need not be rational with respect to what is desired now.

But ultimately this mechanism is still a psychological property of the person and can only "come to the surface," so to speak, by manifesting itself within the conscious relationship between the subject and the object. In this sense, Freud actually maintains the usual direct linear connection between subject and object, except that this connection is now subdivided into conscious and unconscious paths.

Rational desires pertain to the conscious mind, while irrational desires can originate in the unconscious mind. For various irrational modes of behavior, Freud posited a corresponding unconscious mechanism—a death instinct, an insecurity complex, an Oedipus complex . . . The genius and depth of Freud's work goes well beyond this simplistic description, of course, but the basic linear structure of his model of desire is accurate enough. Despite the influence of Freud's work, many psychologists have come to understand that this repertoire of complexes, while providing a tool for psychological analysis, does not really *explain* human behavior.

An alternative and extremely powerful model is offered by mimetic theory, originally developed as a theory of literary criticism by the French thinker René Girard.[3] Girard recognized that the insights of the most profound novelists and playwrights—in particular Dostoyevsky, Stendhal, Shakespeare, Cervantes, Proust, and Flaubert—have one crucial feature in common when describing human relationships. Relationships between desiring subjects and their objects of desire always involve *another person*, and this other person acts as a model for the desire of the subject. The subject essentially imitates the desire of the model toward the object. Girard calls this "mimetic desire," alluding to the ancient Greek concept of *mimesis*, with its more complex and violent connotations than the word "imitation." A person's desire, therefore, is not simply his or her own desire, nor is it driven solely by some inherent property of desirability within the object itself. It is always a function of another person's (or other people's) desire. In other words, human desire is not a linear relationship between subject and object—it is a *triangular* relationship between subject, object, and model.

We are of course discussing here those desires that are characteristic of *human* beings, as opposed to the sort of appetitive or instinctual desires common to all animals.[4] If I desire a drink of water, it may be that I simply haven't had a drink in a while and I'm thirsty. No model is required for me to feel thirsty. However, if I turn my nose up at Aquafina and insist on drinking only Perrier, one might rightly begin to suspect some external influence that has little to do with my actual thirst. The triangular structure of desire has in fact always been intuitively understood by the advertising industry. Commercials do not ask us what we like or want, they *tell* us what we like and want. They define themselves as models, often using popular celebrities, so that we may pattern our desires along the same lines. Any parent of a young child will recognize the mimetic nature of desire immediately. A child may be completely disinterested in a certain toy until another child picks it up, at which point he may cry as if he had desired nothing but that toy from the day he was born. This child's desire is clearly in imitation of the other child.

And speaking of parents and children, it is worth returning for a moment to Freud, to highlight the important difference between his model of desire and that of mimetic theory. Freud's model does indeed involve "triangular" relationships, in particular family triangles between father, mother, and child. But as the psychiatrist Guy Lefort says to René Girard in *Things Hidden since the Foundation of the World*, "Freud invents the Oedipus complex as a way of explaining all these triangles." Which means, of course, that Freud does *not* use triangular desire to explain the Oedipus complex. To quote Girard in the same chapter of the book:

Mimetic desire and the Oedipus complex are incompatible for two reasons. (1) For Freud, the desire for the mother as object is an intrinsic one; there can be no question of it being based on something else, let alone another form of desire. . . . If desire for the "nurturing female" is original, natural and spontaneous, it cannot be derived or copied from anything else at all. (2) For Freud, the father certainly serves, from the son's point of view, as a model for identification. . . . But this model for identification is never a model for desire.

In other words, Freud's notion of desire is still something directed from the subject toward an object of desire, something "intrinsically human" directed for one reason toward the mother and for another reason against the father. This is what I mean when I refer to an "object-oriented" or "linear" model of desire.

The crux of Girard's work has been to show that all human desire is triangular, so that there is ultimately no desire without a model. In fact, it is the connection between subject and model, not subject and object, that ultimately dominates the dynamics of human relationships. Because both subject and model have their desires directed toward the same object, sooner or later they tend to find themselves in competition for that object. The model becomes simultaneously a model for desire and a rival for the object of desire. Subject and model become entangled in what Girard calls a "double bind" (borrowing the term from Gregory Bateson), in which each person imitates the other's desire, thereby intensifying the rivalry between them. The situation escalates under a form of positive feedback, in which imitation fuels rivalry, which drives stronger imitation, which fuels more intense rivalry . . . Beyond a certain point, the rivalry becomes the dominant leg of the triangular relationship, making the object itself less and less important, and eventually even forgotten altogether. This is the classic "Hatfield and McCoy" scenario, in which two rivals are so intent on fighting each other that they have forgotten what they are fighting about. It is also the fundamental logic underlying all forms of mob violence and persecution.

The disappearance of the object of desire is a key aspect of mimetic theory. For our purposes, however, it is also important to note a related aspect of the logic of mimetic desire: the apparent disappearance of the *model*. In this case we are not referring to an actual absence of the model, as can be the case with the object, but to the fact that the model can be hidden within the complex web of human relationships, or even within the psychology of the subject himself. Human desire must always have a model, and in the absence of any external model desire will even imitate itself. This is in fact the dynamic behind the phenomenon of obsession, which is essentially a desire that feeds on itself and drives itself to a frenzy—a desire that desires for the sake of desire. The triangular configuration of object-subject-model is a minimal description of the relational nature of human desire. The key insight, however, is that this is a collective phenomenon involving multiple people, as opposed to an independent autonomous connection between the subject and the object. This has led Girard to replace the concept of individuality with that of "interdividuality," expressing the fundamentally relational nature inherent in each person.

This collective dimension opens the door to a variety of ways in which the relation between subject and model can manifest itself. The most profound and interesting effects of mimetic desire are found, not only between two people, but between communities, between societal classes, between a person and the collective understanding of cultural institutions. In the brief example given earlier about preferring Perrier to Aquafina, where exactly is the model? One cannot point to a specific person that I am imitating. And yet it is intuitively obvious that this desire to drink only Perrier is defined by something external, something independent of either my thirst or the quality of the water itself (since Aquafina would satisfy my thirst just as well).

To be a human being is to be part of a culture, and a culture is essentially a fabric of models—a sphere of influence that transcends the subject in the same way that "humanity" transcends each human.[5] This understanding of mimetic desire on an anthropological level makes it sometimes difficult to trace the precise connections between subject and model. And yet a growing number of Girardian scholars are making a career of doing just that.

• • •

Ultimately, the central contribution of mimetic theory to our understanding of human relationships and human culture is its redefinition of the structure of human desire. The contradictions and paradoxes that seem to define the "human condition," and the inadequacy of so many attempts to explain them, can largely be traced back to the simplistic assumption that human desire is characterized by a direct relation between the desiring person and the object of desire. In the context of this linear relationship, human behavior often appears nonsensical or arbitrary—the product of a kind of temporary insanity or a quasi-magical unconscious mechanism.

We have seen how both modern and classical science rely on a similar understanding of measurement or observation. Like the classical vision of desire, the act of observation is considered a direct linear relationship between the observer and the object of observation. Like Dostoyevsky's underground man, the measurement problem reveals the difficulties embedded within this linear relationship. The most convincing "explanation" for these difficulties involves an arbitrary mechanism of collapse that, like a Freudian complex, does not really explain anything. These parallels between the fundamental problems of quantum theory and mimetic theory reveal a profound formal similarity.

If the problems have the same form, is it not reasonable, then, to suggest that their solutions may also have the same form? This may seem like a bold suggestion if by "solution" we imagine a sort of closed-form mathematical answer to a quantum-mechanical equation. But Girard did not "solve the problem of desire" in this way. What he did was to recognize a fallacy in the way we were thinking about desire, and ultimately he brought us back to an understanding that is philosophically, theologically, and empirically consistent. Similarly, quantum mechanics is not wrong in itself, and we already know how to solve quantum mechanical problems. But, as the first two chapters show, there are symptoms of a fallacy or inconsistency in our thinking. Can we leverage the parallels between mimetic theory and quantum theory to help us understand what quantum mechanics is trying to tell us?

As a minimum, the mimetic nature of the human observer

forces us to reexamine the thread connecting this particular subject to the object. As soon as we combine the notion that the object is a function of the subject (as implied by the measurement problem) with the insight that the human subject is inherently interdividual (as suggested by Girard), the situation becomes more complex. At any point in time, it is obvious that a group of people will generally have a wider knowledge base than any single person. Two people looking in different directions will observe something more than either person alone. In this sense, it is a tautology to say that what *I* know is different from what *we* know. But if what I know affects the physics of the system in front of me, then it follows that the way that system actually is to *me* is different from how it actually is to *us*.

This simple deduction is significant, because the interdividual nature of the human observer ensures that every "me" is in reality an "us." The difference between "my world" and "our world" is not only a difference between separate individuals, it is a difference that is somehow embedded within *each* individual. How does one draw a simple straight line to this type of self? The philosophical problem of objectivity versus subjectivity, which lies at the heart of the measurement problem, must be reformulated to take into account the subjective relationship *between* people as well as their *collective* relation to the objective world.

Following the logic of mimetic theory, then, let us postulate a different structure for the relationship between observer and observed—between subject and object. Let us assume that this relationship is not linear but *triangular*. This means that any observation or measurement that a person makes is not solely a function of the inherent measurability or measurable properties of the object of observation. It means that behind every observation there is the implication of another observer. In other words, every observer requires a model, and because of this every observation is a function of another person's observation.

It is not that one observer affects or changes the object itself, thereby affecting another observer's measurement. This would simply describe the standard model with two observers, each with a linear connection to the object. The secondary relation between the observers, if there is any, could then be derived from each one's relation to the object.

This is in fact the current understanding of the EPR effect. One observer's independent relationship to his object affects the state of that object, which then determines the state of another object that may or may not be measured by another observer afterward. Such a model forces us to postulate a "spooky" nonlocal connection between localized objects. As we pointed out in chapter 1, an alternate connection between the observers themselves would alleviate the strangeness of the situation, and a triangular model of observation provides this connection via the inherently nonlocal character of the interdividual observer. According to this model, the crucial relationship is not the one between subject and object, but the one between subject and subject (i.e., subject and model), and it is this relationship that affects the connection of each to the physical system. This difference in structure is shown graphically in figure 2.

The traditional structure of observation involves two subjects (observers), each connected to the object via direct observation. The two observations O_1 and O_2 are completely independent of each other, and subject 1 has no necessary connection to subject 2. In the triangular model,

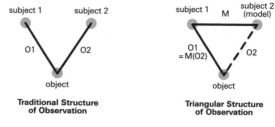

**Traditional Structure
of Observation**

**Triangular Structure
of Observation**

Figure 2.

there is a connection M between the two subjects such that subject 2 becomes a model for subject 1. This means that O_1 is no longer independent of O_2—they are related through the mimetic connection M, so that O_1 is a function of O_2, represented here as $M(O_2)$. The dashed line O_2 reflects the fact that this connection between the model and the object need not be a simple direct observation. Note that if we think of the connections O_1 and O_2 as relations of desire rather than observation (call them D_1 and D_2), we have exactly the model of human desire described in mimetic theory. The structures of observation and desire are identical, and we need only replace terminology to move from one context to the other.

We should be careful to make a subtle but important distinction here. I am not claiming that observation is itself a form of human desire, or that the connections between observers are connections of desire per se. My strategy is simply to ask, "What if observing subjects are related to each other in the same sort of imitative way as Girard's desiring subjects?" The first implication of this is that the observer must be a human being, because that is the only thing that can enter into a mimetic relationship. But this means that the observing subject and the desiring subject are one and the same, which in turn implies that mimetic observation and mimetic desire are two sides of the same coin—two dimensions or manifestations of the fundamental mimesis that defines our interdividual humanity.

The task, then, is not to link observation to desire, but to link observation to mimesis in a more general sense. But of course, once this linkage is made, the corresponding linkage to human desire is unavoidable, because the mimesis that defines the human observer is the same mimesis that defines the human desirer. Consequently, the logic of mimetic desire is *the* model one must use to develop the corresponding logic, and resultant implications, of triangular observation.

Right off the bat, these implications have an impact on the quantum mechanical concepts of uncertainty and probability, which refer ultimately to the multiple potential relationships between the observer and the object of observation. The quantum theorist, working within the limitations of a linear model of observation and driven by the modern obsession with "objectivity," looks only toward the object for explanations. If the wave function reveals a kind of schizophrenia in our observations, he insists that it is the object that is schizophrenic, and then he marvels at the almost human fickleness of the physical world. If the EPR effect reveals an underlying interconnectedness between observations, he insists that the observers themselves have nothing to do with it, and then he wonders how this purely physical connection could possibly work. A triangular model of observation allows us to put uncertainty back where it belongs, where it makes sense: in the human person. The physical world can be thought of as completely objective in itself and inherently nonprobabilistic. It is only when the physical world is related to a human observer that a subjective element is introduced, and this element is probabilistic because it is not merely individually subjective but *interdividually* subjective. That is, quantum probability derives from the contingent relationship between the subject and his or her models, as a characteristic of the inherently interdividual nature of the human observer. It is not derived from multiple objects inherent in the object, but from multiple subjects inherent in the subject.

From a strictly practical or mathematical point of view, we can still think of the wave function in probabilistic terms, and use it to determine the most likely outcome of a particular measurement or observation. There is therefore no reason to think that the results would not agree with those of traditional quantum mechanics. Recent efforts in the field of psychology to model human cognition and decision-making, mentioned briefly in chapter 2, testify to this. The objective of these efforts is precisely to apply the mathematics of quantum mechanics to human intentional behavior. It may be that this work will eventually lead to a formal mathematical foundation for triangular observation as well.

The *collapse* of the wave function, however, must be understood in a fundamentally different way. Unlike the probabilistic object in traditional quantum theory, the interdividual observer does not suddenly cease to be interdividual after an observation is made. And yet there is a discontinuous change, a sudden transition from multiple potential observations to a single actual observation. This fact of measurement cannot be avoided, and it is indeed a measurement problem as long as we look through the narrow scope of our linear observational model to the object itself.

If we remember the human observer, however, and the human act of observation, we will notice immediately that one aspect of this discontinuity, namely the fact that it is discontinuous, reflects merely the transition from potential knowledge to actual knowledge discussed in the last chapter—what Bernard d'Espagnat calls "the mysterious transition from the potential to the actual."[6] But while this intentional act of the intellect explains the transition from before to after, it does not by itself explain the transition from multiple to single. For Aquinas, of course, there is no collapse of anything. Nothing is lost. It involves rather the *addition* of knowledge to the object. For quantum theory, however, there is indeed a loss—the loss of all states of the object in superposition except one. Triangular observation effectively brings these two models together by taking into consideration the inderdividuality of the observer.

Every object that is not actually known is potentially known by an observer. But to be known by an interdividual observer is to be known by multiple intellects—or, more correctly, by an intellect that is a function of multiple intellects. Using Pieper's metaphor of the "grasping mind," we can say that, before an observation is made, there are many hands reaching for the object, each with some potential to grasp it. But in the end only one hand can *actually* grasp it. The object can only *actually* be one thing, and an individual person cannot observe an object as more than one thing, or from multiple perspectives simultaneously.

We therefore can speak of a sort of "collapse" in triangular observation, but not a collapse from the indeterminate object to the determinate object. It is rather a collapse from the interdividual to the individual. But even this is not really a collapse in the quantum mechanical sense, because to be interdividual means precisely to be *both* collective *and* individual. It is indeed I, Pablo Bandera, who struggle always to possess the object of my models' desires and exert my individuality over and against the influence of my models. I have no one to blame for this but myself. And yet my efforts are always disappointing to some degree, because in making someone my rival I have already acknowledged him as my model. I cannot help but be myself, but in the end I am not merely myself. The transition from potential to actual, therefore, involves a reduction but not a loss. It is the shift from the collective dimension of observation, which is inherent in each observer, to the individual dimension of observation, which is also inherent in each observer.

To the scientific reader, this dual nature of interdividuality may seem reminiscent of the well-known particle/wave duality observed in the laboratory. This is the strange ability of nature to act both as a particle, a localized point or "bundle" of energy, and a wave with a distributed wavefront of energy. Light waves exhibit diffraction patterns when shone through small openings, but light photons interact with individual atoms and molecules as if they were miniature billiard balls. And objects such as subatomic particles have the same characteristic: emitting a beam of electrons through a small opening results in a diffraction pattern, just like a light wave. Whether nature acts as a particle or as a wave depends on how specific measurements are taken during the experiment—in other words, on how nature is observed.

One could conceivably draw an analogy between this particle/wave duality and the individual/collective duality of the interdividual person. Except that whereas this duality in physical nature is shocking (or was, until we grew to accept it), the notion of an interdividual person is psychologically and philosophically reasonable. It may be subtle and profound, but not fundamentally disturbing or hard to believe. Similarly, the collapse of the wave function is somewhat "ridiculous" (in Schrödinger's words) when understood as an arbitrary trick of the physical object, but it falls quite naturally out of a triangular understanding of human observation. In the final analysis, this "collapse" or transition from potential to actual is not very mysterious or magical, and it is only as miraculous as the fact that we human beings are capable of knowing anything at all.

If this is true, then it provides a way to reconcile the probabilistic physics of quantum theory with the realism of Thomistic ontology, because it means that the form of the object is indeed *the one* form of the object, as Aquinas insists. The multiple forms implied by quantum theory are a range of multiple observations, which is, after all, what the wave function is meant to represent. But if this multiplicity is a reflection of human interdividuality, then it allows the object itself to remain ontologically singular. We can therefore understand the potency of Thomistic philosophy and the potency of quantum theory as two aspects of the observer-observed relationship. The former refers to the knowability of the object, the latter to the uncertainty of the observer. In this way the two approaches are not only compatible but complementary.

Moreover, this understanding of the wave function satisfies the condition of scientific completeness. The interdividuality of the human observer does not refer to a number of points of view that may or may not exist, but to a range of perspectives that make up the actual complete perspective of the mimetic observer. Just as the state of the object was thought of as a range of states, the "state" of the human observer is a range of states defined by the relationship to his or her models.

At first glance, it may seem that a triangular structure of observation may be difficult to defend when external models are not readily identifiable. Exactly the same criticism was made early on against mimetic theory as well. And the same question can be asked of quantum theory today regarding the EPR effect: If an object is physically separated and considered an isolated system, where is this other object that is supposedly responsible for determining its state? To what other object can it point if it is all alone? This notion of being able to point to a model is misleading. It is important to resist the temptation to imagine two or three people staring at an object and somehow affecting each other's individual measurements as if by telekinesis. Such an image actually attempts to force the concept of a subject-model relationship into the old linear model of observation, with individual subject-object relationships and a secondary relationship between observers occurring in the background. The communication between observers need not be so simplistic, and must be understood as an integral part of the complete structure of observation.

As with mimetic theory, the central point is not that one observer imitates another observer, but that the process of human observation, like that of desire, is a *collective* and *mimetic* phenomenon. It is not always obvious who the models of our desires are. Yet we see that even when the model is not physically present, when there is in fact nothing to which one can physically point, even when the subject is completely alone with his or her desire, the triangular structure of that desire remains. Similarly, triangular observation must be considered within the broader context of the relationship between the subject and the object, that is, between the subjective and the objective.

The philosophy of science has been struggling for a while now with this fundamental relationship, and I am not the first person to try to relate science, or a scientific phenomenon, to social or collective processes. The last few decades in particular have seen the development of a "sociology of scientific knowledge," represented in different ways in the work of Thomas Kuhn, David Bloor, Imre Lakatos, Bruno

Latour, and others.[7] But while there is no question that other philosophical theories of science involve notions of sociality, or collectivity, or collaboration, or relativity, or group paradigms, this should not prevent us from recognizing the unique significance of a triangular model of observation. When Girard first started developing mimetic theory, he was not the only person talking about the "Other," or the imitative qualities of man, or the social dynamics underlying the individual. And yet his concepts of mimetic desire and interdividuality have had an enormous impact on everything from cultural anthropology to systematic theology. This didn't happen because Girard "discovered" or "invented" imitation or collectivity, but because he restructured these into a model of desire that brought out their *universal* significance.

Despite the common misconception in grade school science, Isaac Newton did not "discover gravity"; he discovered and defined the universality of the law of gravity, as something that determines the motions of every massive object from nearby apples to distant planets. Einstein was not the first to postulate a quantized form of energy, but he borrowed Planck's idea and applied the concept to light, converting it from a strange property of blackbodies into a universal characteristic of nature. In a similar way, Girard discovered the universality of the law of human desire. What is this universal law, which separates us from the other animals and underlies our fundamental humanness? Mimesis—the fact that we are not merely ourselves but are also, anthropologically, a product of the people with whom we share this world. I, in my turn, am suggesting that we apply this universal law of humanity to the human observer, and in so doing reinterpret the strange features of nature revealed by modern physics.

If one wants to judge the originality of this idea by comparing it to recent developments in the philosophy of science, I suggest it may be more fruitful to go further back, to the wisdom of the ancient Greeks. When Narcissus looked at the world and saw his reflection, he mistook it for something, or rather someone, in the world itself, independent of himself. When modern physicists look at the world, they make, I think, a similar mistake. They see what is in fact a reflection of themselves and mistake it for something in the world itself, not a person but a feature of the thing they are looking at. And just as Narcissus succumbed to narcissism, physicists tend to succumb to physicalism.

Of course, Narcissus merely looked at the surface of the world. The image on the water reflected his surface—his handsome face—and his love was equally superficial. But what if he could have looked deeper, beyond the shallow waves of the water to the unreachable wave function of nature? He would have been looking, not at the water, but at the waterness of the water, at that which made the water what it was. The image there would have been a reflection of what made Narcissus what *he* was—not merely the image of a human being but an image of his humanity. It is in this invisible mirror of the quantum world where we see a reflection of the mimetic dynamics that make us who we are—the image of our interdividual selves, distorted and moving among uncertain waves.

4
People and Things Hiding in Plain Sight

Recent developments in the philosophy of science, and in particular what has come to be known as the sociology of scientific knowledge, would seem to have a lot in common with the hypothesis of triangular observation. There is indeed a certain affinity between some of this work and the ideas proposed in this monograph, at least in attitude if not in content. It is worth taking a closer look, then, to see where they overlap and where they don't. From there we can delve more deeply into some specific interpretations of quantum theory that also, in different ways, appear somewhat "triangular" on the surface.

The sociology of scientific knowledge is generally divided into two broad schools of thought, referred to as the weak program and the strong program. The former tries to understand the relationship of social and psychological factors to the workings of science as an objective study of nature. Accordingly, it sees these factors primarily as sources of error or deviation from the objective norms and standards of good science. The strong program, by contrast, insists that science is properly understood by taking into account all relevant factors, social and nonsocial, regardless of whether or not they can be said to have led to error. This is clearly closer to the spirit of my hypothesis. But in both cases it should be understood that these theories do not generally have any direct connection to the metaphysical issues of quantum theory described in the first chapter. The sociology of scientific knowledge deals mainly with the practice of science and the process of scientific development, and how these are to be understood within an overall framework of scientific knowledge. As David Bloor, a leading figure in the strong program, explains: "The aim isn't to explain nature, but to explain shared beliefs about nature. The enquiry is into the character and causes of knowledge, or what passes as knowledge, and not (in general) into the objects which the knowledge is meant to be about."[1]

Even Bruno Latour, who criticizes other philosophies of science for not being radical enough in their restructuring of the subject-object relation, characterizes his own work as follows: "As a working definition, therefore, it could be said that we are concerned with *social* construction of scientific knowledge in so far as this draws attention to the *process* by which scientists make sense of their observations."[2]

The specific human dynamics that underlie this process, namely the intersubjective relationships in which sociologists of scientific knowledge are so interested, would probably be best understood using the insights of mimetic theory. One day, perhaps, some astute philosopher of science will take on this task and make important contributions to the field. In the meantime, the hypothesis of triangular observation attempts to apply these insights to the ontological questions raised by quantum theory, to the workings of nature at the quantum level. My goal is to relate, not just human activity to science or scientific objects, but humanness to physicality.

And yet, despite this categorical difference, it is hard to resist the parallels between certain aspects of the sociology of scientific knowledge and a triangular model of observation.

Bruno Latour's work, for example, is full of provocative phrases like "the mediation of the laboratory" and "the co-production of collective things" that would capture the attention of a scientist familiar with mimetic theory. Latour has said explicitly that he wants to perform an anthropological study of modern scientific culture in the same way that classical anthropologists study archaic and primitive cultures. What could be more sympathetic to the work of René Girard, who has effectively shown how to do just that?

But a closer examination of these parallels shows them diverging quite early on. The "anthropology" of Latour is done in the context of scientists performing experimental research and developing their respective fields of study. A central part of his philosophy is the idea that all the discussions, debates, and proposals that ultimately define the state of a scientific field are worked out in an "agnostic field" of ideas and statements, and in principle this field is indifferent to any claims of objectivity or truth. Scientific statements get thrown into this field and are modified and pared down according to how they relate to other statements in the field. The statements that survive in the long run are those that require the greatest effort or cost to modify, and these become the defining statements of science. It is a field of competition, but between statements, not people. That is, there is competition, but there is no rivalry. And this is ultimately because there is no model from which a rivalry can emerge. In other words, any concept of mimesis is entirely absent. Latour notes that "the notion of agnostic contrasts significantly with the view that scientists are somehow concerned with 'nature,'" because they are really primarily concerned with making and modifying statements in the agnostic field, and these only later, independently of any objective physical reality, reify into the physical objects of scientific practice (laboratory equipment, documentation, technology, etc.). In the case of a triangular model of observation, one could similarly (but not quite accurately) say that scientists are not primarily concerned with nature. But this would be because they are more concerned with each other, with their models rather than their objects, and there is nothing agnostic about it.

The strong program, as described by David Bloor in his arguments against the philosophy of Latour, is actually more aligned with a triangular model of observation. Leveraging the work of his colleague Barry Barnes, Bloor notes the "self-referring" character of social institutions. That is, a social institution exists primarily as that which is agreed upon and referred to as such by the members of society. The reference to the institution constitutes the reality of it as an institution. In this way, the distinction between subject (the thing referring) and object (the thing referred to) is blurred, and Bloor sees this as a possible mechanism to relate psychosocial factors to the external natural world. "The interesting theoretical task is to combine this model of a social institution with the sociological insight that *all* knowledge has the character of a social institution. And this includes, of course, knowledge of an independent reality, with an independent object."[3]

This by itself sounds very similar to a triangular model of observation, which also connects real external objects to a type of social relationship. But Bloor's goal here is not to define the nature of this social dimension, but on the contrary to highlight the fundamental "materialism" of the strong program. Our knowledge of the objective world is indeed determined by social factors, but the underlying subject-object relationship must be understood in "individualistic" terms.

> Consider an organism learning about its environment by causally interacting with it. . . . It is an active process in which one part of nature (the subject) interacts with another part (the object). These fundamental, individual causal and biological processes do not, of course, derive from culture but are, rather, presupposed by culture. . . . Thinking about how a computer might learn to interact with a simple environment of movable and shaped blocks would be one way of bringing the basic and pre-given relation of subject and object into view.[4]

This is clearly another expression of what I have called the "linear" relationship between the subject and object, taken for granted here as it is almost everywhere else. The picture we get, then, is that of a group of individuals with individual observations of nature, but the "tool" that each person uses to understand his or her observation is the set of rules and conventions that he or she shares with the rest of society. "It is only by collectively sustaining a set of concepts that genuine and coherent reference to an external reality becomes possible."[5]

Sociologies of scientific knowledge, therefore, not only maintain a linear understanding of the basic subject-object relationship, they also rely on a higher-level logic of social cooperation that runs counter to the mimetic dynamics of a triangular model of observation. In the latter, people do not merely share certain social or cultural conventions with their neighbors; their relationships to nature are *imitated* from their models. The difference between sharing and imitating is the difference between cooperation and rivalry. And this difference stands out clearly when we look more closely at the kind of social processes sociologists of science have in mind. Probably the most famous example, and one referred to explicitly by Bloor in the above arguments, is that of Thomas Kuhn's scientific "paradigms." In the course of doing "normal science," Kuhn explains that working within a particular paradigm will seldom evoke overt disagreement over fundamentals. Men whose research is based on shared paradigms are committed to the same rules and standards for scientific practice. That commitment and the apparent consensus it produces are prerequisites for normal science, i.e., for the genesis and continuation of a particular research tradition.[6]

In other words, as long as people look at things through the same eyes, through the lens of a particular paradigm, they proceed along happily in agreement. But this consensus can be disrupted, Kuhn says, when two scientists look at something through *different* eyes, leading to revolutions within or between different scientific paradigms. In a triangular model of observation, the focus is precisely on the reverse of this dynamic: it is when people look at things through the same eyes, that is through the eyes of their models, that we have a tendency toward conflict, not consensus. The probability of conflict and rivalry is greater when things become more the *same*, not when they become different. If a revolution occurs, it will be as a culmination of this escalating "undifferentiation" within the group.

The point is not that Kuhn was wrong. There are certainly social and cultural institutions in which people share certain norms and conventions. But these institutions, which have developed in various complex ways over time, have their roots in the fundamental anthropological mechanisms that gave birth to culture in the first place. These are precisely the mechanisms of mimetic desire and mimetic crisis described by mimetic theory, and cultural institutions developed primarily as a way to stabilize society against these mimetic forces. They are still at work beneath the surface, so to speak, of social and cultural institutions, determining the relationships of people to each other and to those institutions. What I am proposing in the present study is that they are also at work in determining the relationships—the quantum mechanical relationships—of those people to the physical world around them.

• • •

Turning our attention now more specifically to philosophies of quantum theory, one might think that the special emphasis on the observer in modern physics would have led to something like a triangular model of observation sooner. Indeed, among the many interpretations proposed over the last century, the ones taken most seriously exhibit features that could be called triangular. But upon closer inspection they all fall back on a linear understanding of the subject-object relationship, and this is ultimately their downfall.

Already in the Copenhagen school, triangular structures can be seen, for example, in Bohr's concept of "complementarity," which comes very close to a triangular model of observation before being sacrificed to the idol of objectivity. Complementarity attempts to establish the completeness of quantum theory by taking into account both the observer and the observed as parts of the complete system. It stems from a basic understanding of scientific measurement as a mode of communication between observers. "The argument is simply that by the word 'experiment' we refer to a situation where we can tell others what we have done and what we have learned."[7]

In order to do this, we must speak in some intelligible common language, which in turn must reflect an intelligible picture of nature. We rely on (classical) concepts such as causality and determinism to provide this intelligibility. Therefore, any experiment we perform, by virtue of its function of communicating to us an experience of nature, can only be understood in terms of the macroscopic images and concepts of classical physics.

But the actual workings of nature need not follow these rules, and in fact have shown that they do not. When experimental measurements drill down to the most elemental components of physical matter, its "quantum nature" begins to manifest itself. This nature defies the usual deterministic descriptions, and so from our own macroscopic perspective nature strikes us as a contradiction, a breaker of its own rules.

Because of this we must make a sharp distinction between what is being measured and what is doing the measuring. These two components of any experiment are fundamentally different in nature. But the point is not to treat them separately but to bring them together as complementary aspects of the experiment. Bohr insisted that quantum theory provides a scientifically complete description of nature as long as both observer and observed are taken into account in this complementary way.

Bohr saw a parallel between this and the "complementary" aspects of the human person. A person can be described in terms of his or her physical or biological attributes, which appear subject to the laws of determinism, or in terms of the mind or spirit, which often appear to defy deterministic rules. A complete description of the human person, however, requires that both aspects be considered together. Such connections between the mind and the body, between the observer and the observed, would seem to suggest a connection between the subjectivity of the observer and the objectivity of the physical world. But Bohr warned against this temptation:

> The decisive point is that in neither case [relativity theory or quantum theory] does the appropriate widening of our conceptual framework imply any appeal to the observing subject, which would hinder unambiguous communication of experience. . . . In complementary description all subjectivity is

avoided by proper attention to the circumstances required for the well-defined use of elementary physical concepts.[8]

Needless to say, if an appeal to the observing subject is "inappropriate," any subjective relation to another observer is doubly so. Bohr therefore brings all the points of the triangle together but refuses to connect them. He insists on including both object and observer in the complete system, and even recognizes the role of communication with other observers as the defining characteristic of any experiment. But the system is essentially static: a collection of parts, in which the function of each is merely to be what the other parts are not. What is missing from this and every other major model of quantum physics is the inherently dynamic character of the human observer. What is missing is the model.

The first major response to the Copenhagen interpretation was hidden variables theory, which essentially tries to create a triangular structure between the subject, the object, and some other unknown unobserved object. Hidden variables theories attempt to get around the implications of the measurement problem by suggesting that quantum mechanics as understood by the Copenhagen school is in fact incomplete. If a certain measurement yields results that appear indeterminate, it is only because we are not accounting for all the factors involved. There must be some other parameter, some "hidden variable," that is acting behind the scenes to cause all these problems. But this triangle is a broken one, with three points but only two sides. There is no connection between the subject and the unobserved object, and of course there is no model for the subject at all. At best, hidden variables can only fill the analogous role of a Freudian complex—a mechanism acting under the observable radar that affects the linear relation between subject and object. In order for the situation to be truly triangular in the mimetic sense, we must posit the existence, not of another object, but of another subject: not hidden variables but *hidden observers*.

On the surface, the term "hidden observers" may seem to suggest something like Hugh Everett's many-worlds theory, with its many hidden observers observing many objects. This interpretation avoids the mechanism of collapse by giving every possibility contained in the wave function a place to exist. Once a measurement is made and a particular state is observed, the other states that were part of the wave function—i.e., that were in "superposition" with the observed state—do not suddenly cease to exist, they simply exist somewhere else, effectively in another universe. Consequently there are many, perhaps infinitely many, universes in which all the states of every observed system (and all the corresponding "versions" of every observer) continue to exist and follow the determinate laws of nature.[9] It imposes a kind of consistency by insisting that the entire universe, including all observers, remains in superposition, in an overarching universal wave function, and therefore never collapses. But this, of course, is not triangular either. Many-worlds theory and its derivatives describe parallel linear connections between subjects and objects, and it is in fact crucial to these theories that these connections remain independent of each other. In contrast, a triangular interpretation hinges on the interaction between multiple observers.

One particularly interesting variation on the many-worlds theory, proposed by David Albert and Barry Loewer, essentially replaces Everett's infinitude of worlds with an infinitude of *minds*. Albert and Loewer point out what they consider to be a crucial problem with many-worlds theory: its central claim that we all remain in quantum superposition even after an observation is made is incompatible with our basic experience that our minds are not perceived to be in any sort of superposition. We never perceive ourselves observing something in more than one state at the same time (how could we?). "The heart of the problem is that the way we conceive of mental states, beliefs, memories, etc., it simply makes no sense to speak of such states or of a mind as being in a superposition."[10]

Note that this is only a problem if the observer is a self-conscious person, that is, someone with a mind and the capacity for introspection. But because of this it calls into question many-worlds theory in general: either the theory works for all observers or it doesn't work at all. This incompatibility between the superposition of physical states and the identity of human self-perception leads Albert and Loewer to the conclusion that a person's mental "belief states" must be separated from his or her physical "brain states." Moreover, these two types of states must be independent of each other, as a person's unique belief state can have no effect on the continued superposition of his or her brain states. This would seem to suggest that each observation involves a superposition of "mindless brains," with only one brain possessing a mind with the unique belief state of the observer.

Albert and Loewer rightly recognize the dubious implications of such a scenario, and one would think that at this point a triangular model of observation might suggest itself, once the significance of the nonphysical mind of the observer is appreciated. But this distinction between mind and brain is not based on an appreciation for the role of the human mind; it is derived from an attempt to account for our experience of the quantum mechanical nature of physical reality. In the end, "many-minds theory" postulates the existence of an infinite number of minds associated with each brain state. These minds are not in superposition—they are separate independent minds that effectively line up with the multiple states of the wave function of the physical brain. This is essentially the opposite of the triangular observation model, which postulates a "superposition of minds" (i.e., the mimetic interrelation of the minds of models and subjects) associated with a single brain.

This difference is made especially apparent in Albert and Loewer's insistence that their theory should be taken seriously for its "realism." By this they mean that the entire physical universe can be modeled by a single universal wave function, without any dependence on the mind or consciousness of a subjective observer. In fact, the mind has become so disassociated from the physical world that the belief states of the observer are, by the logic of the theory, necessarily wrong. A belief state will always think an object is in state 1 or state 2, when in fact it is always in a superposition of multiple states. What sort of realism is this, in which our knowledge of reality has nothing to do with reality? Thomas Aquinas would be appalled.

But Albert and Loewer go further. They note that this multiplicity of minds and brains does not guarantee that the same mind will be associated with the same brain over time. This in turn calls into question the very notion of an actual person's identity in the first place. "Perhaps, however, it could be argued . . . that the conception of a mind persisting through time is an illusion . . . Could we go further and eliminate all reference to minds?"[11] Many-minds theory, therefore, starts out by recognizing that each observer has a mind and ends by suggesting that each mind has nothing to do with the observer. Surely a triangular model of observation is preferable to this type of circularity.

Probably the closest interpretation of quantum theory to a triangular structure of observation is the one proposed by Carlo Rovelli in his paper entitled "Relational Quantum Mechanics."[12] Rovelli does indeed posit a possible relation between different observers of the same object. According to this theory, an observation is only significant as a measurement relative to a particular observer, in the same sense as that described by different reference frames in relativity theory. Just as reference frames can be related to each other via the appropriate mathematical transformation, so can quantum mechanical observers be related to each other.

The focus, however, remains on the connection between the subject and the object, the main thesis being that this connection reflects a relative state of physical reality rather than an absolute one. The connection to another observer, while it can be related mathematically, is essentially independent. A second observer may acquire some information about the first observer only because it follows the first one in time, and therefore can measure the result of the first measurement. But the second observer does not act as a model in any way for the first observer.

•   •   •

The difficulty in picturing a truly triangular structure of observation is not due to any lack of imagination. It has deep philosophical roots embedded in a dualistic vision of the world that was effectively canonized in Western thought by René Descartes. With the famous conclusion "cogito ergo sum," the inner self was set up as the true ground of all knowledge, in opposition to the unreliable outer world of physical objects and space. This "self" is utterly individual, believing only in itself and skeptical of the existence of anyone or anything else. Western philosophy for the last three hundred years has largely been an effort to understand the relation between the "inner" self and the "outer" world.

The philosopher Roger Scruton describes the Cartesian man in the following way: "In considering the relation between thought and reality he is considering not 'our thought' but 'my thought.' Maybe there are no other people: maybe only I exist, and what I take for other people are no more than paintings on the wall."[13] This position is basically taken for granted in the physical sciences, which have always been careful to distinguish the subjective observer from the objective world. The act of measurement or observation is significant only insofar as it relates to the particular observer, and this observer is essentially alone in the world. Any other observer is quite literally treated as a part of the external landscape. In quantum theory this is taken to the extreme, as the observer himself is considered part of the physical landscape. Nowhere is this dualistic attitude more clearly reflected than in the measurement problem itself, which is essentially a kind of allergic reaction to the mixing of the subjective observer with the object of observation.

The radically materialistic world envisioned by Thomas Hobbes played a significant role in the formation of the modern scientific mind as well. But whereas Descartes finds the ground of epistemological certainty in the ultimate autonomy of the ego, Hobbes sees this autonomy as leading to an inescapable *uncertainty*. For the Hobbesian man, the utter subjectivity of the individual, who knows the world only through physical experience, reduces him to a purely material being responding to the physical laws of nature and causality. In contrast to Descartes's distrust of the body, Hobbes insists that the *only* reality of which we can be certain is the body, without which it would be impossible for the mind to even be aware of the external world or itself in it. This places a hard limit on what is knowable at all, as all knowledge is mediated by the sensory inputs of the body, which are mere subjective images of real physical reality.

And yet, despite this emphasis on the external world, the Hobbesian man is no less individualistic than the Cartesian man. The existence of this world is entirely derived from the purely subjective experience of the individual. As the philosopher Stephen Gardner has shown, Hobbes does not reach his "materialist" or "empiricist" conclusions by opposing the "rationalist" method of Descartes, but rather by taking the radical subjectivity of the Cartesian individual to its logical extreme:

What later thinkers call the "objectivating" or "reifying" manner of science is in fact a strict consequence of an irreducible perspectivalism. It reflects the point of view of an individual who looks on the world as "external," that is, precisely from within, enmeshed in an uncertain world of resistances, obstacles and threats.[14]

And so Western thought, and especially scientific thought, remained dependent on an underlying image of the internal individual versus the external collective. In the mid-twentieth century, however, there was a fundamental shift in philosophy, the repercussions of which have yet to be felt deeply in the scientific world. The decisive refutation of the Cartesian "theory of mind" finally came from Ludwig Wittgenstein in his *Philosophical Investigations*. Wittgenstein showed that the Achilles' heel in the phrase "I think therefore I am" is the word "I." The mistake is in thinking that this "I" is an independent or irreducible reality. How can we even speak of an "I" unless it is in relation to the Other? Again, Roger Scruton sums up Wittgenstein's central conclusion well:

If you can think about your thinking, then you must do so in a publicly intelligible discourse. In which case, you must be part of some "public realm," accessible to others. This public realm is also an objective realm. . . . Moreover, we must reject the Cartesian picture of the mind, which derives entirely from a study of the first-person case—a study of what is revealed to me, as I cease to meditate on the "external world," and turn my attention "inwards." We must recognize the priority of the third-person case, which sees the mind from outside, as we see the minds of others.[15]

It is not difficult to see how this parallels a triangular understanding of humanity and human relationships. We reach Wittgenstein's basic conclusion by simply replacing Descartes's purely individual self with a self that is a function of other selves. This assumption destroys any possibility of knowledge as being grounded in the isolated self, and forces this ground out into the real objective world of other people.

I have no doubt that a detailed study of the development of epistemology from a triangular perspective would be fascinating, but this is obviously beyond the scope of this monograph. Our task here is merely to recognize a key factor in the shift away from the Cartesian theory of mind. The connection between the subjective person and the objective world, which was problematic at best for Descartes, was achieved by introducing a collective dimension to the very foundation of the problem. The only way to understand each self is to consider his or her relation to other selves. Moreover, the only way to understand the relation of each self to the objective world (which we call observation or measurement) is to consider his or her relation to other selves. When we ask, "How does the world look to me?" we are in fact asking, "How does the world look to *us*?" In short, the hypothesis of triangular observation is entirely consistent with the most profound and convincing conclusions of modern philosophical thought. If it seems strange to our "scientific" minds it is only because of our reluctance to let go of the egocentric view of the world that seemed so natural to Descartes.

For the modern physicist who wants to place all physical causes outside the subjective person and in the "objective" outside world, the suggestion that quantum uncertainty is derived from human observers will of course be a point of contention. For his benefit, then, we may point out that triangular observation does indeed suggest that the source of uncertainty is in the outside world. But the outside world is not made up merely of inanimate physical objects, it is also made up of other people, other observers, without whom we could not even speak of an objective point of view. Without these other people any observation would necessarily be "my observation" only. But this situation does not actually exist. To be human is to be interdividual, which implies a relation to other people. There is no such thing as a human being alone in the world, and the hypothesis of triangular observation takes this fundamental fact into account.

As long as the phenomenon of observation is understood as a linear connection between subject and object, any exploration of the measurement problem is extremely limited. We are forced to account for various unexpected results somewhere within the simple and direct relation between the observer and the observed. If we look in the direction of the latter, suggesting that the answer lies within the physical system itself, then we are ultimately led to the random and arbitrary universe of the Copenhagen school and its variants. If we look in the direction of the observer, we impose a quasi-magical element in which the observer somehow arbitrarily changes the laws of nature during a measurement. This eventually leads to the network of metaphysical meanderings in which some thinkers have gotten lost. A triangular structure of observation provides a third dimension in which to work out the measurement problem. The answer lies neither solely in nature nor in the "consciousness" of the observer himself, but in the relation of both to the model: a relation that defines the triangular structure of observation in general.

If this is indeed the "solution" to the measurement problem, then it has profound implications for the problem itself. It means that the usual distinction between observer and observed, and ultimately the entire worldview that describes nature in terms of subjective and objective, basically misses the point. The distinction is misleading because it completely ignores the model, and it is the model after all that determines our relation to the world. Up to now the measurement problem, and in fact all discussions of scientific observation, have claimed a sort of autonomy either for the observer or for nature. But both extremes derive from a dualism that reduces everything to a single dimension—a line between two points. This limited perspective has made it very difficult to look around and recognize the Other as part of our world: not just another object among others, but another subject, upon which our own subjectivity and objectivity depend. Once this recognition is made, the term "subjective" takes on a much richer meaning, one that refers to our place in the network of relationships that defines our situation as objective observers.

We can see now why the measurement problem has been so difficult to unlock. The key was hiding within the structure of the problem itself, which has always been stated in terms of the opposition between the observer and the observed. And yet this was, in a way, inevitable. It is precisely the recognition of the Other as the ground of our own subjectivity that we as human beings find so difficult. The "romantic lie" of our autonomy (in Girard's words) tends to see the Other, if he is seen at all, as a rival—as part of the problem rather than the solution. It is not surprising at this point that we should once again converge with mimetic theory, for we are describing our situation, not just as observers, but as human beings in general. As Girard says:

> The objective and subjective fallacies are one and the same; both originate in the image which we all have of our own desires. Subjectivisms and objectivisms, romanticisms and realisms, individualisms and scientisms, idealisms and positivisms appear to be in opposition but are secretly in agreement to conceal the presence of the mediator.... They all defend the same illusion of autonomy to which modern man is passionately devoted.[16]

5
The Roots of Reality

If there is one obvious message we can take away from the last hundred years of development in physics it is that the world, and in particular the physical world, is fundamentally different from how we thought it was. Quantum theory has directly challenged our understanding of determinacy and causality—those things that have traditionally allowed us to recognize the real world as real. In the words of Niels Bohr: "In every field of experience we must retain a sharp distinction between the observer and the content of the observations, but we must realize that the discovery of the quantum of action has thrown new light on the very foundation of the description of nature."[1]

It has indeed thrown new light, but it is a harsh light, full of shadows and dark corners. The philosophical confusion and metaphysical hand-waving described in the first few chapters suggests that we are still far from a clear and complete understanding of the world. The point is not that we are entirely on the wrong track, but we are clearly missing something important. Thomas Nagel expresses a similar concern when discussing the various arguments surrounding the mind-body problem. "I believe it is already clear that any correct theory of the relation between mind and body would radically transform our overall conception of the world and would require a new understanding of the phenomena now thought of as physical."[2]

The relation between mind and body is fundamentally the relation between the subjective and the objective. It is precisely on this field where modern physics has thrown down the gauntlet. The fact that the observer is not merely an observer but also a participant in the workings of an objective event forces us to consider what this means when that observer is a subjective human being (even if that represents only a special case). How do we reconcile the implications of this with an "objective" view of nature? What is the right way to translate the observer-observed relationship into the subjective-objective relationship, and can this be done without reducing objective reality to a figment of someone's imagination? Any valid model of reality must be able to answer these questions.

Albert Einstein understood this, and did his best to address these issues. In his book *Relativity*, Einstein uses the abstract concepts of space and time to distinguish between an "experience" and an "event." Space and time are effectively ordering principles that allow us to organize a set of subjective experiences into an objective event. It is especially interesting that, in moving from the subjective to the objective realm, Einstein requires the introduction of *another* observer.

> A person *A* ("I") has the experience "it is lightning." At the same time the person *A* also experiences such a behavior of the person *B* as brings the behavior of *B* into relation with his own experience.... For the person *A* the idea arises that other persons also participate in the experience.... In this way arises the interpretation

that "it is lightning," which originally entered into the consciousness as an "experience," is now also interpreted as an (objective) "event." It is just the sum total of all events that we mean when we speak of the "real external world."[3]

We can think of an "event" as effectively a measurement of a particular experience—an interaction between a person and nature that relates an experience of nature to the "real external world." Einstein, then, posits a *collective* dimension behind this measurement. In other words, the relation between the subjective and the objective is grounded in the collective dimension of measurement. This is essentially a condensed description of triangular observation.

And yet Einstein never actually explores the full implications of his intuition, and stops short of postulating a truly triangular structure of measurement or observation. Despite a recognition of the real existence of objects behind the original experience, as well as the real existence of "person *B*," the collective dimension of the experience-event relation is internalized within the psychology of person *A*.

The concepts of space, time and event can be put psychologically into relation with experiences. Considered logically, they are free creations of the human intelligence, tools of thought which are to serve the purpose of bringing experiences into relation with each other.[4]

This is something closer to what one might call a Freudian model of reality. One is tempted to speak of an "ordering instinct" or an "experience complex." Rather than acknowledge the influence of the other person on the objective reality of an event, the relation between observers is reduced to a psychological mechanism by which personal experiences are generalized—projected like Platonic shadows onto the walls of the external world, but never actually affecting the world in reality.

Ultimately Einstein remains trapped within the usual dualism that divides the world into inner (subjective) reality and outer (objective) reality. One can interpret events subjectively or analyze experiences objectively, but these are separate and independent operations. The implication is that the inner reality is not really real, or at least not objectively so, and it therefore occupies a secondary role in the study of the physical sciences.

Thomas Nagel has a different approach. As mentioned earlier, he insists on the obvious reality of human subjectivity, and considers its exclusion from most models of nature to be a fundamental problem in science and philosophy. In contrast to Einstein's analysis, Nagel says the following:

> If we try to understand experience from an objective viewpoint that is distinct from that of the subject of the experience, then even if we continue to credit its perspectival nature, we will not be able to grasp its most specific qualities unless we can imagine them subjectively. We will not know exactly how scrambled eggs taste to a cockroach even if we develop a detailed objective phenomenology of the cockroach sense of taste.[5]

Any model of objective nature must therefore include subjective reality, which leads Nagel to distinguish between the two in a subtly different way. If a subjective point of view is that of certain individuals, then the purely subjective is defined as the point of view of the single individual. Objectivity, therefore, is defined as the degree to which we are removed from this individual point of view. Unlike the usual static definition of objective reality as whatever lies outside the human mind (which is usually considered synonymous with physical reality), Nagel defines it as a sort of dynamic process, a movement away from subjectivity, which is a movement away from the individual.

Whereas Einstein sees subjectivity and objectivity as two independent realities, Nagel sees them as two parts of one reality, or rather the two extremes of a spectrum of reality. But these two parts are in competition with each other. They grudgingly share the same claim to reality, but at any point we can ask how objective or subjective a certain point of view is compared to another. This is clearly still a form of dualism, but one that at least tries to see itself for what it is. In fact, Nagel's main objective is not to eliminate the subjective/objective dualism but to level the playing field for both and point out their inherent limitations.

The main limitation stems from the following logical characteristic of objective reality: it is ultimately derived from a subjective basis. As any realist will agree, the value of objectivity is that it brings us closer to the real existence of things (what one might call the "truth" of things) by transcending the subjective individual point of view. But this quality of transcendence—this ability to observe one's point of view from outside, to watch one's own watching, to think about thinking—is not inherent to objective nature. On the contrary, it is the exclusive quality of the subjective person. Objectivity is achieved, therefore, through a progression of essentially subjective operations that, paradoxically, lead away from the purely subjective individual point of view.

Nagel's point is that any complete model of objective reality should not forget the subjective basis from which it comes. But this makes it difficult to actually achieve a purely objective point of view, as any objectifying step must always include its subjective basis.

The idea of objectivity thus seems to undermine itself. The aim is to form a conception of reality which includes ourselves and our view of things among its objects, but it seems that whatever forms the conception will not be included by it. It seems to follow that the most objective view we can achieve will have to rest on an unexamined subjective base.[6]

The tension between subjectivity and objectivity becomes especially acute in Nagel's study of human autonomy. Human actions are real events in the world and are therefore observable from an objective, external point of view. We can certainly understand the physical manifestations of a human action in terms of objective physical causes and effects, but what really determines a human action in the first place is the person's decision to do it. A decision is a special sort of "cause." As discussed earlier, there is no necessary cause for deciding to do one thing and not another, in the same way that the collapse of the wave function has no physical cause "pushing" it toward one state or another. A free human action occurs ultimately for no reason other than that was what the person decided to do. That is what makes it free. It is by nature irreducibly subjective, and any attempt to view it objectively essentially turns it into something it's not. At first glance it seems to be a step in the right direction: viewing a human action objectively promises to show it for what it is and reveal the true causes behind it. Placing human action on the spectrum of objective causes and effects, as one natural object among others, should give us control of these actions and leave behind the capricious idiosyncrasies of human freedom. But we quickly find that this is a lie, because taking control of freedom means destroying it.

But the real problem, according to Nagel, is that no one is really convinced by this strategy. Human action is, as we've said, irreducibly subjective, which means purely individual. And purely individual for Nagel means strictly autonomous. Every individual intuitively believes in the autonomy of his or her actions and decisions. The threat of objectification, therefore, is not merely a logical problem but a crisis of belief. In this Nagel takes a very practical and intuitive attitude. All objective theories of mind proposed so far fall short because they all fail to address the basic conviction that our actions are indeed ours. "Nor is it possible to simply dissolve our unanalyzed sense of autonomy and responsibility. It is something we can't get rid of, either in relation to ourselves or in relation to others. We are apparently condemned to want something impossible."[7]

While Nagel struggles with the dualism he rightly criticizes, we see that something else is needed, an approach that truly integrates the subjective and objective into a single model of reality. Such an approach is taken, for example, by Hannah Arendt in one of her few purely philosophical works, *The Life of the Mind*. In this long and detailed analysis, Arendt begins by describing the usual assumption behind the philosophical dualism we've been discussing—namely that what matters is the object itself, while the appearance of that object to an observer is secondary. She then proposes inverting this hierarchy, suggesting that it is the appearance that we should take seriously as that which determines the objective reality of any object. "That appearance always demands spectators and thus implies an at least potential recognition and acknowledgement has far-reaching consequences for what we, appearing beings in a world of appearances, understand by reality, our own as well as that of the world."[8]

In contrast to the independent realities of Einstein and the competing realities of Nagel, Arendt speaks of the objective reality of subjective appearances. An object does not merely prove its existence by appearing to people, it exists *so that* it can appear to people. Subjective and objective reality imply each other. Consequently, the dilemma that is so problematic for Nagel essentially forms the philosophical basis for Arendt's thought.

In fact, Arendt's thought owes much to Thomas Aquinas, as she herself acknowledges. What she describes in terms of "appearances" is essentially what Aquinas describes in terms of "knowledge"—all things exist for the sake of being observed or known. But if this is true, then we must recall the problem raised by quantum theory in chapter 2. Unless we fall into a kind of philosophical relativism, Arendt's reality of appearance would seem to imply a one-to-one correspondence between appearance and object. Not that an object cannot appear differently to different people, but its appearance to each person should be unique (as it is, for example, for each reference frame in relativity theory) if it is to correspond to the true object. Quantum mechanics, however, suggests that the same object corresponds to multiple potential appearances, each one no less real than the other. Is the reality of appearance, then, more accurately described as the "realities of appearances?" Are we thrown back into the disturbing world of the Copenhagen school?

We are indeed, unless we consider the interdividuality of the observer for whom the object appears. We showed in chapter 3 how we can recover the full power of Aquinas's (and Arendt's) insights by attributing the different appearances of an object not to the object itself but to the different perspectives inherent in every interdividual human observer. If we consider the wave function as a reflection of the subject-model relationship in a triangular model of observation (rather than the subject-object relationship in a linear model), we can reconcile the results of quantum mechanics with the implications of Thomistic philosophy.

We can also now address the problems raised by Nagel regarding the tension between subjective and objective reality. This tension exists because of the way he sets collective objectivity in opposition to individual subjectivity, as extremes defining a sort of spectrum of reality. One begins with the pure subjectivity of the autonomous individual and progresses asymptotically toward the goal of pure objectivity by including a more and more collective point of view. What Nagel never seems to suspect is that there is no such thing as an autonomous individual in this sense—a singular independent perspective that defines a purely subjective reality. Even the most isolated individual person is a product of, and conditioned by, the mimetic relationships that define him as a human being in the first place. The illusion of autonomy is precisely the "romantic lie" exposed by mimetic theory. Consequently, every human observer, by virtue of his or her interdividuality, is always already objective to some degree. In other words, while there is a subjective basis to objective reality, there is also an objective basis to subjective reality. Nagel does not only move toward an unachievable goal of pure objectivity, he moves away from a false ideal of pure subjectivity.

Whereas this could be called a sort of blind spot for Nagel, he nonetheless has the intuitive depth to recognize the existential nature of the issue. He describes the problem of free will, which for Nagel is basically the problem of the loss of individual autonomy, as "a loss of confidence, conviction or equilibrium." "Just as the basic problem of epistemology is not whether we can be *said to know* things, but lies rather in the loss of belief and the invasion of doubt, so the problem of free will lies in the erosion of interpersonal attitudes and of the sense of autonomy."[9]

Nagel sees this anxiety as a logical result of the tug-of-war between the subjective and the objective. But it would be more correct to say that it is a reflection of the tug-of-war between the self and the other, the struggle within each one of us to assert our own autonomy over and against the influence of our mimetic models. This is the fundamental conflict at the heart of human culture, and it is indeed a crisis of "confidence, conviction or equilibrium," among other things.

In the end, the paradox noticed by Nagel—the more *objectively* responsible we are for our free actions the less autonomously free, and therefore less responsible, we actually are—does not ever go away. But neither is it merely a logical wall we run up against in our pursuit of objectivity. Nagel's intuition goes to the right end but comes to the wrong conclusion. For while this tension between subjective and objective, between freedom and responsibility, is an insoluble dilemma for the purely autonomous self, it falls naturally out of an interdividual understanding of the self. As mimetic creatures, our motivations and actions are not merely our own, and yet we are ultimately responsible for our actions. This is precisely the paradoxical nature of human freedom, expressed succinctly by Jesus when he pleaded on the cross, "Father, forgive them; for they do not know what they are doing."[10] We are to some degree driven unwittingly by our mimetic relationship with our models, to some degree a victim of contagion. Yet we need to be forgiven—we are to some degree responsible. There is no way to know where to draw this line between self and other, between culpability and what Girard calls "méconnaisance," and this is a problem for Nagel (and most others) who is trying to place subjectivity on one side of the line and objectivity on the other. But Nagel does not notice that there is another line, the one connecting the subject to the model. This line does not separate subject from object but rather *connects* the two, through the model, in such a way that incorporates the broad concepts of "objectivity" and "subjectivity" into a single triangular structure of reality.

• • •

Einstein, Nagel, and Arendt each have different approaches to understanding reality and come to different conclusions. All three, however, are unabashed realists. They are not interested in the suggestion that the world around us is actually some sort of illusion, or that reality is something that exists only in our minds. They all believe in the real "objective" existence of the object in front of them. It is interesting, then, that for all three the concept of objectivity refers in some way to the degree to which multiple people participate in the experience of an object or event. Collectivity is the measure of objectivity.

This in fact works against the understanding of some sociologies of scientific knowledge, despite their emphasis on collective social forces and their effects on the scientific world. Latour in particular sees a practical equivalence between these social forces and the objective reality or "truth" of science. We have been working toward an understanding of reality that avoids a stagnant dualism by incorporating the subjective and objective "perspectives" (in Nagel's terms) into a more integrated and holistic model of the world that takes *both* perspectives seriously. But Latour asks, since social processes and objective reality are effectively interchangeable anyway, why even bother talking about the social at all? "By demonstrating its pervasive applicability, the social study of science has rendered 'social' devoid of any meaning."

Needless to say, if "social" has no meaning, then "human" has even less. This is diametrically opposed to the philosophical position behind a triangular model of observation. Moreover, this radical wiping out of the distinction between subjective and objective makes objective reality essentially irrelevant as well, which is not at all what Einstein, Nagel, Arendt, or Aquinas had in mind. Referring to the competing scientific statements in his "agnostic field," Latour says, "The set of statements considered too costly to modify constitute what is referred to as reality. Scientific activity is not 'about nature,' it is a fierce fight to construct reality."

If we were to translate this to the metaphysical level of quantum theoretical interpretation, it would be something closer to Wigner than to Einstein or Aquinas. The latter were *very* concerned with nature, the objective reality of nature, and the place of human beings within nature.

What I have tried to stress in the present work is that, while collectivity is the measure of objectivity, it is also the measure of the human—to be human is to be part of humanity. As Arendt says, "Men, not man, inhabit this planet."[11] Mimetic theory further refines this notion by showing that it applies, not just to the population of the world, but to each individual person by virtue of his or her interdividuality. As mimetic creatures, we carry within each one of us that collective dimension that relates us to each other and to the objective world around us.

Each human being, therefore, is both individual and collective (i.e., interdividual), which implies a sort of confluence of subjectivity and objectivity. The medium that binds these two realities into one is mimesis, and the point at which they intersect is the moment of human observation. Aristotle and Aquinas already understood this moment as an act of (subjective) appropriation of the (objective) form or essence of the object. Quantum theory says that objective physical reality is a function of objective observation. Einstein, Nagel, and Arendt say that an objective observation is a collective observation. Mimetic theory says that the collective dimension inherent in any human observation is a function of mimesis. To state all this more succinctly, actualized objective reality implies objective observation, which implies collectivity, which implies mimesis. Consequently, there is a *metaphysical* relationship between human mimesis and what we consider the objective reality of the physical world around us.

If this is true, then it leads to a rather obvious question: Was the physical world fundamentally different before the existence of human observers? Or, put another way, did the birth of humanity somehow change the nature of physical reality itself?

If we answer too quickly in the affirmative, we run the risk of heading toward the extreme reached by Wigner in his reinterpretation of Schrödinger's famous thought experiment. Wigner recognized the crucial role played by human consciousness in the measurement problem, and this led him to conclude that the physical world did not really exist, or existed in a state of ontological limbo, until the first conscious observer brought it into concrete existence by collapsing the wave function. But our question is not about the effect that any one person had, as if by magic, on nature. It does not refer to any ability that people have to manipulate nature via the power of consciousness. The question refers rather to the relationship between the objective world of physical reality and the (equally real) subjective world of human observers: not to the effect of people on nature, but to the place of people in nature.

The first humans looked at the world in a fundamentally different way than their prehuman ancestors, a way that connected them, not just to the existence of things, but to the nature of their relationship to those things. Girard describes this as the "first non-instinctual attention," in which the purely appetitive desires of prehuman animals transformed into the mimetic desires of humans. One Girardian scholar describes the moment of hominization as the transition from the immanent to the transcendent.[12] A threshold was crossed that placed humanity irrevocably beyond the realm of purely appetitive desires, beyond the realm of stagnant causal relationships, beyond the realm of immanent physical reality.

We ceased to be individual creatures driven by unilateral needs and appetites and became "interdividual" creatures motivated by our relationships with each other. To be human is to be mimetic.

Triangular observation essentially describes the other side of this phenomenon. When the first humans looked at the world in a fundamentally different way, the world responded in a fundamentally different way. In a sense, the laws of physics were allowed to manifest themselves "honestly," in accordance with the truth of their objective relationship to an intersubjective humanity. For the first time, nature was not merely looked at, it was *observed*, and at that moment quantum mechanics was born.

There is no reason to imagine that the physical world does not exist in and of itself, and unless we are willing to accept a God that "plays dice" with the universe, we must assume that the laws of physics are consistent. By "consistent" we mean that the laws of physics are the same for each observer, in each observer's local reference frame: what Einstein called the principle of relativity. But relativity also shows how a consistent set of laws for each individual observer can look different when compared from a more global perspective—that is, a nonlocal perspective from which one can compare the results of an experiment in one reference frame to those of another reference frame. For example, an object that is moving relative to a stationary observer may appear stationary to a moving observer. If any one observer had access to this nonlocal perspective, he or she could conceivably observe these differences in behavior, which would appear as inconsistencies within his or her own local reference frame. It was because of this that Einstein rejected any such notion of a nonlocal perspective. And yet the EPR effect suggests that this nonlocality does indeed exist.

It cannot be physical nature alone that connects itself to this nonlocal perspective, since the laws of physics can operate at any point in time only on a local level. Any physical system is, by virtue of its finitude, a localized system. This is what makes the EPR effect so paradoxical. The evidence shows that the physical universe is nonlocal, but there is nothing purely physical that can account for this. If the manifestation of a particular law of physics is to be affected by a nonlocal perspective, this can only be via the observer, and more specifically via those properties of the observer that allow him to transcend the local parameters of the physical system: namely, human self-awareness and intersubjectivity. Consequently, while the laws of physics are consistent in themselves, how these laws manifest themselves in the presence of humanity can be very different. It is our existence in this world *as observers* that makes the difference. Our question about physical reality before humanity, therefore, has implications, not so much for nature, but for us as intersubjective human beings.

To help us understand this, let us imagine a two-dimensional world in which all physical reality exists on the surface of a flat sheet of paper. There exist the concepts of length and area, but there is no notion of volume. Now suppose some event occurs (it doesn't matter precisely what) that opens up a third dimension for this world. What sort of strange new things would the inhabitant of this world observe? A square, when viewed at certain angles, would appear to have unequal sides. Circles would sometimes look like ovals, and straight-looking paths would actually be curved. Yet the laws of two-dimensional geometry are still valid in three dimensions. This new three-dimensional world would look very different, not because the laws of the two-dimensional world would have changed, but because those same laws would manifest themselves in different ways and produce new results depending on how they were observed.

In the same way, the presence of humanity opens up a new dimension in which the classical laws of physics can operate. We can call this the dimension of intersubjectivity, or more correctly the dimension of transcendence, because it is what allows us as human beings to see, not just the world around us, but ourselves within the world and our relationship to it. This transcendence reveals itself in our relationships to each other through mimetic desire. It reveals itself in our relationship to the physical world through quantum mechanics. In either case, it exists only because of the presence of mimetic human observers.

Now, while a two-dimensional projection can be generated from a three-dimensional object, the reverse is not possible: one cannot use shapes in two dimensions to create a solid object in three dimensions. Similarly, the classical laws of physics can be derived as special cases of the more general laws of quantum mechanics, but quantum mechanics does not follow at all from classical physics. There is nothing in classical physics that would suggest quantum mechanical behavior. A specifically unpredictable observation was required (namely the ultraviolet catastrophe) in order to point us in the direction of quantum theory. If this difference between the old physics and the new physics were merely epistemological, the implications would be fairly innocuous: we could simply say that the world has always been this way, and we just haven't noticed it until recently.[13] However, if this difference is tied ontologically to the existence of human observers, as we suggest here, then the implications are more profound. Quantum mechanics is not merely a different understanding of the old physics, it is a new manifestation of the old physics made possible by the presence of human observers. Consequently, the fact that we cannot derive modern physics from classical physics means that we cannot derive the existence of humanity from the classical laws of physics. Nor can we derive it from the laws of modern physics because these laws are made possible by the existence of humanity in the first place. There is no way to relate the behavior of physical laws alone to those in the presence of human observers without the a priori assumption of the existence of humanity.

This poses a serious problem for any scientific theory that would treat the human being as a purely physical or biological entity. Even without considering such things as the moral or spiritual qualities of mankind, humanity's physical relation to nature makes it irreducible to nature. No scientific theory can move continuously (or "naturally") from the laws of physics to the existence of humanity. The birth of humanity did more than open a door for nature, it was itself the opening of a door that was not there at all until it opened.

6
From Reality to Truth

The search for objectivity is ultimately the search for certainty—certainty that the object in front of me is in actual fact that object and is in actual fact in front of me. It is an attempt to escape the uncertainty of the subjective point of view, to replace mere opinion with ontologically independent fact. But, as Nagel discovered, there is an asymptotic limit to this approach. No matter how increasingly objective one's point of view, it is still a point of view—it always must include the subjective basis from which it is derived. Consequently, while uncertainty in the world around us may decrease with increasing objectivity, it can never disappear altogether. There is a minimum amount of uncertainty inherent in objective reality.

Once again, Nagel's description of reality mirrors that

of quantum theory. Heisenberg's uncertainty principle claims a similar minimum uncertainty at the heart of objective reality. But of course, once again, Nagel's description relies on the inclusion of the human. His uncertainty derives from a trade-off between subjective and objective reality. Heisenberg's uncertainty appears at first glance to be a trade-off between two objective realities, such as position and momentum or energy and time. But it is in fact a trade-off between their *uncertainties*, between knowledge of one and knowledge of the other. While Nagel shows uncertainty to be a subjective limitation on objective reality, Heisenberg shows uncertainty to be an objective limitation on subjective knowledge. Taking the two together, we begin to get a picture of reality in which the lines between subjective and objective are very blurry. Uncertainty does not merely point a finger at the subjective observer or at the objective world, but points beyond both. It says there is something more than what can be defined either subjectively or objectively. In other words, there is an ontological limit to what we call reality.

But what does this mean? We seem to be saying that there is a point at which reality ceases to be real. What sort of "reality" exists beyond the limit of reality? The world we see around us seems real enough. Does this mean that the world we don't see is not only uncertain but somehow unreal? This is in fact how some quantum theorists have come to understand things. Heisenberg himself seemed to reach this conclusion almost reluctantly after noting the parallel between quantum probability and Aristotle's concept of potentia. He couldn't quite bring himself to say that the world was not real, but at the quantum level it seems to be somehow "not as real" as we would like.

> In the experiments about atomic events we have to do with things and facts, with phenomena that are just as real as any phenomena in daily life. But the atoms or the elementary particles themselves are not as real; they form a world of potentialities or possibilities rather than one of things or facts.[1]

Wolfgang Smith has taken this further to develop a detailed interpretation of quantum theory based on different levels of reality. More specifically, he distinguishes between the "corporeal" world of objects that can be observed directly and the "physical" world of attributes that cannot be observed, only measured using a corporeal object such as a ruler or thermometer. For example, one does not observe $10\degree$F, one simply feels cold. One does not observe six feet of tallness, one sees a tall man. In order to see that this tallness is six feet, one needs to use something like a tape measure to reflect this physical attribute onto the corporeal world. Smith points out that these two worlds are not the same, and maintains that modern scientists have mistakenly confused one for the other.

Like Heisenberg, Smith sees a correlation between the probabilistic wave function and Aristotle's potentia. But the distinction between physical and corporeal allows him to be more specific: he associates Aristotle's potentia with the physical world—the unobservable world of measurable quantities. These quantities, these "physical objects," are purely mathematical and therefore not real, at least not in the same sense as the corporeal objects we see and feel. But neither are they merely logical or philosophical constructs. Smith refers to them as "existentiated mathematical forms," which suggests that the physical object is in some sense *there*. And in fact, the corporeal world of observable things cannot exist without the corresponding world of physical attributes. So the physical object is there, but it is somehow there without being real; it is "existentiated" but doesn't quite "exist."

> However, we should remind ourselves that physical objects prove ultimately to be nothing more nor less than certain "potencies" in relation to the corporeal world. It is by no means unreasonable, therefore, to surmise that existence, properly so called, "begins" on the corporeal plane.[2]

Smith insists that all the strangeness of quantum theory can be swept away by reestablishing the crucial difference between the physical and corporeal worlds, and by recognizing that only the latter is real.

> The moment one forgets that this so-called [physical] universe constitutes but a sub-existential domain—a mere potency in relation to the

corporeal—one has created a monster. For indeed, the physical domain, thus "hypostatized," becomes forthwith the prime usurper of reality, the great illusion from which a host of baneful errors spring. It is neither a small nor a harmless thing "to lose one's grip on reality!"[3]

It may be philosophically tenable to speak of different forms or levels of reality, but the notion of an objective reality that isn't real is problematic. It rightly made Heisenberg uncomfortable, and it can lead to a confused and unconvincing metaphysics, such as the ontological limbo proposed by Wigner. While Smith cautions us not to lose our grip on reality, his insistence on the existence of a "sub-existential domain" gives the impression that we have done just that.

Roger Paul uses a different but similar strategy when he associates the potentialities of the wave function with the accidental forms of the object. In doing this he believes he is divorcing the wave function from the substantial object itself, which allows him to claim that the wave function does not really exist. But even if we admit that these accidental forms cannot exist by themselves as independent things, we still can only speak of them in relation to a real actual substance, which ties it ultimately to the reality of the object. It may be difficult to pinpoint a specific spot within the wave function and say, "It is there," but the wave function itself still refers to an "it," to something that is in some sense "there" and therefore exists.

This is not merely intuitive, it is more consistent with Aquinas's understanding of potentia in the first place. In his study of Aquinas's metaphysics and its relevance to various fields of science, Gerard Verschuuren reminds us that the concept of potentia is in fact associated with matter rather than form, and matter is what determines the primary being of the object. The nature of prima materia is to provide the potential for change, and this change is actualized in the form of a specific thing. It is the form that changes, but it can only do this because its existence is grounded in the being of the prima materia.

Prime matter is only and always found with some substantial form that gives it determination. Yet, although it never exists in and of itself, prime matter is real; it is a real principle of change. It is the real potency that every material thing has in order to undergo an indefinite amount of change.[4]

It would indeed simplify matters if the range of states represented by the wave function didn't really exist. In fact, it does not seem very controversial at all to say that an object that only *potentially* exists does not, by definition, *actually* exist. But if by this we are implying that the object is free to violate certain principles of physics or metaphysics because, after all, it doesn't exist, then we are taking the word "potential" too far. "Potential" here does not mean the potential to exist *at all*, it refers to the existence of the object in potentia. It is the potential for the object to exist in a certain form, and therefore to be observed or known, which implies a preexisting object in the first place. The object has this potential because it has being, grounded in the prima materia of that object. Its potential to be what it is is derived from the fact that it is. One cannot simply separate potentia from being. Objects exist in potency or in act, and even if we distinguish between these two types of existence, we must acknowledge that they both do in some sense exist, not merely hypothetically but in reality. In this context it is worth recalling what Joseph Pieper said earlier regarding the object in potentia:

> It is the mind that changes an *already existing but only potential relationship* between objective reality and subjective cognition into actual fact. . . . a mutual correlation between *being* and *mind exists already* prior to any actual perception. The object-subject relationship, in traditional ontology, is not at all "a creation on the part of the mind"; *it precedes any activity of the mind.*[5]

Moreover, the scientific completeness of quantum theory insists that each one of the states of the wave function corresponds to the real object itself, as opposed to a merely hypothetical object or an object that could exist in principle. For example, one can imagine that an object is in a particular state but could have in principle been in another state. A person who hasn't observed the object yet could very rightly say that, from his perspective, the object has the potential to be in either of the two states, and he can't know in which state he will observe it until he actually observes it. But in this case the object in fact has no potential to be in the other state at all. We are dealing here merely with an epistemological limitation of the observer. For Aquinas, potentia refers not to the knowledge of the observer but to the knowability of the object. This other purely hypothetical potential we have just described may or may not have anything to do with the real world. Quantum mechanics, however, does have implications for the real world. The whole problem with quantum theory is that its strangeness manifests itself in the real observable world—its reality imposes itself on us whether we like it or not.

We cannot honestly avoid talking about the "reality" of the wave function, or its ontological implications, which is why quantum theorists, despite themselves, keep talking about it. And despite the various confusions and complications, the basic intuition of Heisenberg, Smith, Paul, and others is surely correct: the correlation between the probabilistic wave function and the concept of potentia seems crucial to suggesting some way of describing the wave function in ontologically palatable terms. But we must go beyond Aristotle and a bit deeper into Aquinas, into this subtle "correlation between being and mind" that determines the nature of an observation. It is there, in an important distinction that Aquinas makes (or rather doesn't make), that we may find the key to our dilemma.

The act of observation is the correlation of the form of an object with that same form in the mind of the observer. The mind, however, can participate in this action in two ways: receptively or creatively. The receptive intellect is something closer to what classical physics has always assumed: it receives the form of an object that is already there.

The creative intellect refers to the active creation of the object's form in the mind of the observer first. In this case we are indeed talking about the essence of an object that has the potential to exist outside the mind, but does not actually exist as it has no matter: it is known *what* the object is but not *that* it is. The will of the person is required to bring the object into existence. For example, a statue exists because a sculptor created it, and he or she created it because the form or essence of that statue existed first in the sculptor's mind. While both the receptive and the creative intellects are used to actualize an object in potentia, we can see that only the latter provides a basis for the existence of the object in the first place. If we are going to talk about the existence or nonexistence of the object in potentia, we must refer specifically to that object in relation to the creative intellect. "The forms conceived by the receptive mind are the foundation of knowing only, while the forms conceived by the active mind are the foundations of both knowing and being."[6]

All that exists is ordered, either potentially or actually, toward a knowing mind, and this correlation between object and mind can be either receptive or creative, and only the latter allows for existence in the first place. Consequently, any object that exists in the real world can only do so as the product of a creative mind. It is clearly not human minds that created the totality of forms that exist and have existed throughout the universe. Aquinas, therefore, reaches the conclusion that all things ultimately exist because of their original creation in the creative intellect of God. This imposes a strong condition on any physical object, in that its existence in the real observable world—its being—implies a direct correlation with the divine intellect. This agreement between what the object is and what God intends it to be is what Aquinas calls the "truth" of the object, so that being and truth essentially imply each other.

But while "the true is a conformity of thing and intellect,"[7] there are two types of intellect, and therefore two ways of talking about the truth. Conformity with the divine intellect is the primary form of truth and the source of being for all things. Without the divine intellect, nothing would exist, and even if "by an impossible supposition"[8] it did exist, it could not be considered true. Conformity with the human intellect, however, depends on the existence of things in the first place, and therefore represents a secondary form of truth. Human truth is ultimately derived from divine truth.

Unlike the human intellect, the divine intellect is purely active, since there is nothing that was created outside of God's knowledge whose form He would receive passively. In other words, there is nothing in the universe that is *potentially* known by God. He knows everything *actually*, which means that everything God knows is everything that actually exists. If God's knowledge is purely active, then potential knowledge can only be human knowledge. The important thing about the object in potentia, therefore, is not that it is some sort of nonreality, but that it is a reality that exists only because of the presence of human beings. "Truth is predicated of all things insofar as they actually conform to God's knowledge, or else, as they potentially are able to conform to human knowledge."[9]

We can say, then, that the object in potentia is in a state of existence defined in some sense or to some degree by the human observer. Once this object is observed and becomes an object in act, it enters a state of existence defined ultimately by its correlation with the creative intellect of God, at which point the truth of that object is revealed, so to speak—it is finally known by the human observer. This is all perfectly consistent as long as we assume that there is but one form, one essence, corresponding to any particular thing that exists. The potential knowledge of that object is the potential acquisition of its form by the human mind, which is the same form as that in the creative mind of God.

To observe something is to actually acquire that form and know the object, so that one can only observe something that is really and truly there. Hence Aquinas's well-known statement that "all things are true and cannot be false."[10]

But quantum theory says that the object in potentia is comprised of multiple forms. Only one of these forms can correspond to the true object of the divine intellect—to the object we actually see and feel, the actualized object of the collapsed wave function. The other forms contained within the wave function clearly *cannot* correspond to the divine intellect. Therefore, while our earlier discussion suggests that the forms or states of the wave function are real, we cannot necessarily think of them as *true* in the sense described above. The crucial distinction to be made, then, between the object in act and the object in potentia is not that one is real and the other is not. The crucial distinction is that the object in potentia is real in the limited context of a scientific objectivity that can be demonstrated in the laboratory, while the object in act is real in the fuller sense of a world aligned with the divine truth that is the source and ground of its being in the first place. There is apparently a gap between the human-defined reality of the object in potentia and the truth of the object in act as defined by the creative intellect of God. This is the limitation to which the crooked finger of uncertainty is pointing: not a lack of objective reality, but a lack of divine truth.

Objective reality, as we have seen, is grounded to some degree on epistemology, although it cannot be reduced to epistemology. It refers to the correspondence between the epistemology of the observer and the ontology of the object. This is the lesson of the measurement problem, as well as the dilemma of Nagel. No matter how objective reality is, it is at best an approximation, a generalization of the subjective epistemological foundation on which it is built. Truth, by contrast, is entirely grounded in being. Thus Aquinas agrees with Augustine that "the true is that which is,"[11] because "the true is convertible with being."[12] Truth possesses precisely that ontological basis which objectivity strives for, the essence and existence of the object independent of any observer. Objectivity is an agreement on the existence of an object: it is or is not as we have measured it. Truth is a *judgment* on the existence of that object: it is or is not.

Taking this into account, then, we can understand Aquinas's statement above a little differently and make the following distinction: Actually conforming to God's knowledge is the measure of divine truth, while potentially conforming to human knowledge is the measure of what might be called scientific truth, which is merely the correspondence with objective reality. Aquinas assumes a kind of continuity between these two, as they both refer ultimately to the same form or essence. [13] For Aquinas, an object is true if it exists, if it is part of the real world. Existence, being, reality, and truth all imply each other. Now, armed with the insights and evidence of quantum theory, we can be more precise, at least with respect to things in potentia. The object in act must be true, as it is necessarily known by God, and is ultimately the product of His creative intellect. But the object in potentia, determined as it is by the intellect of the interdividual observer, does not reach this level of divine truth and therefore can involve a kind of lie.

Now, the notion of nature involving a lie is indeed strange and difficult to come by if we understand the multiplicity of forms represented by the wave function as physical states of the object itself, independent of any observer. In particular, it is very difficult to reconcile this with the basic tenet of Thomistic ontology that we have been discussing—namely that truth is "convertible with being," and therefore a lack of truth implies a lack of being. We are led to the conclusion that a lack of truth results not only from a lack of being but from a kind of surplus of being as well—that is, from a surplus of states that make up the being of the object in potentia. What's worse, this lack of truth would apply to all the forms of the object except one (the one that actually gets observed), betraying a sort of impersonal and arbitrary favoritism that is no less disturbing than the measurement problem with which we started.

With a triangular model of observation, however, we find a seamless compatibility with Aquinas's ontology, because the multiplicity of forms represented by the wave function does not represent any superfluous being of the object, but rather the multiple intellects inherently related to the interdividual observer. Unlike the multiplicity of the physical object, the multiplicity of the interdividual person does indeed reflect a lack of being—what Girard calls an "ontological sickness." The fact that a person's desire is always in imitation of a model means that the person wants his model's desire to be his own. But desire is not an object that can be possessed independently of the other person. For the subject to possess his model's desire, to claim that desire as his own and *not* his model's, he must actually *be* his model. This is what the subject really wants; not the object of desire per se but the being of the model that desires it. The subject perceives the model as possessing something the subject lacks. But this something cannot be the physical object itself, which is only the catalyst for imitation in the first place. What the mimetic subject actually lacks is autonomy, the power to call his desires his own, and it is this perceived autonomy and power, this ontological independence, that he envies in his model.

In the case of quantum theory, this ontological sickness is not only the result of a "romantic lie," as Girard calls it, but of a lack of divine truth in the Thomistic sense as well. It is in the mimetic relationship to our models of observation where we discover a distortion of that "correlation between being and mind" which defines the object in potentia. In the end, once we recognize that the problem is not merely a lack of reality or objectivity but a lack of truth, we are led unavoidably to suspect the involvement of a human observer, since the only thing in the world capable of lying in the first place is a human being. Whereas the existence of an object that isn't true is an ontological contradiction, the existence of a lie requires precisely the existence of a human being.

Being, as we have said, is ultimately derived from the divine intellect. For an inanimate object this means simply to have a form that correlates to the creative intellect of God. For the human person, however, made "in the image and likeness" of God, this must involve, at least to some degree, a correlation between the human intellect and the divine intellect. If these two are not completely aligned, it can only be because of the uniquely human quality of freedom, or free will. For this is essentially what it means to be free: to follow one's own will, which may or may not be God's will. As we have said, the human intellect is a function of its mimetic models. But it has the freedom to follow different models. It is this unique power of freedom that allows the specifically human observer to separate the objective reality of his existence from the divine truth of his existence—what he is from what he is supposed to be. What quantum theory does is to show how this ontological gap in the human observer is reflected in the physical world, through the crucial relationship between "being and mind," between the observer and the observed.

What we call "our models of observation" can be understood perhaps a little more clearly now than it was in chapter 3 when we first mentioned it. The difference between potency and act corresponds to the difference between the interdividual self and the individual self. But this also therefore reflects the difference between the observer's relationship to his human models and the observer's relationship to God. To the degree that my observation coincides with that of my human mimetic models, that observation is objectively real. To the degree that my observation coincides with that of God, that observation is true.

· · ·

What I am calling an "ontological gap" is best reflected in the distinction made by mimetic theory between "vertical" transcendence and "horizontal" or "deviated" transcendence. The latter is a sort of pseudotranscendence generated by the mimetic relationships and interactions with our models. The desire of a single person may seem subjectively attached to that person. But a desire that is shared between models and imitated throughout the community takes on a kind of objectivity that transcends the mere individual. To understand this fully we must recall the tendency of mimetic desire to lead to competition and violence.

In imitating the desire of my model, I create a desire for the same object that is desired by my model. This puts me in competition with my model, making him both my model and my rival. These two aspects encourage and feed off each other: the stronger my imitation the more intense the rivalry, which in turn strengthens my imitation further. It is a positive feedback mechanism that increases in intensity and eventually (or often very quickly) leads to violence. As anyone who has witnessed an outbreak of mob violence will agree, this sort of thing is extremely contagious. Violence leads to more violence, and as it escalates it can spread like a plague and consume an entire community of people.

This is what Girard calls a "mimetic crisis." It is characterized by a decreasing level of differentiation between the members of the group, as every person imitates every other person in a frenzy of mimetic violence. Eventually the situation must reach a state of such unanimous and homogeneous violence that only two possibilities remain. The group can spontaneously destroy itself, each person simultaneously killing and being killed in a sort of human supernova. Or, what is more likely and often observed, a single member of the group can stand out for the slightest and most arbitrary reason, drawing the attention and therefore the violence of the crowd onto himself. It may only take one person to notice, to point a finger, and the hyper-mimetic tension binding the crowd together will pull all other eyes and fingers in the same direction. At this point the center of attention becomes the relief valve for the escalating violence that was trapped within the undifferentiated mob. The ease with which this can happen is intuitively understood. Imagine stepping outside your house into the street to find yourself in the middle of a mob riot. Your instinct is to keep your head down, to not draw attention to yourself. If you do, you know that not just some but all eyes may turn on you, and then you're done for.

The choice of victim of this spontaneously focused violence is completely arbitrary. It could be anyone at all, but once that person becomes the center of attention, the rest of the crowd is in instant agreement. The mimetic contagion that locked people in an undifferentiated mob of all against all is the same contagion that bonds them in a mob of all against one. There is no discussion, no looks of approval or nods of agreement, there is only the unanimous and unquestionable conviction that the person everyone is looking at is the problem.

This important detail cannot be overstated. There is no thought process that occurs between the moment the victim is noticed and the moment everyone "agrees" that he or she is the problem. They are essentially the same moment. This is something much deeper than the causal conscious or subconscious relationships of group psychology. The collapse of the crowd onto an arbitrary victim occurs quite spontaneously from within the dynamics of the mimetic crisis, as the result of an essentially stochastic process. My choice of words, of course, is deliberate. We are describing a noncausal dynamic very much like the one that results in the collapse of the wave function onto an arbitrary state of the system.

And just as the collapse of the wave function determines the objective state of a system, the collapse of the crowd onto the victim is in fact a point of maximum objectivity. At this moment the members of the crowd are not really concerned with who is legitimately responsible for the crisis or who started it. They only care that this person is the one to blame. And strictly speaking, from the perspective of everyone involved (except perhaps one), they are not wrong. Once everyone agrees on a victim, that person does indeed become the solution to the problem, the one to blame, precisely because everyone agrees on it. This is evidenced by the very empirical fact that the solution works. As soon as the victim is killed or expelled from the group, and as a direct result of this, peace returns to the group and the crisis is over. According to Girard, this is the underlying mechanism of sacrificial society and, ultimately, human culture. If something works, it must exist in some sense—it must be real. It can't work if it doesn't exist.

But does the victim actually deserve this fate any more than anyone else? Is this victim essentially any different from the rest of the crowd? No, and this is where we can speak of a lack, not of objectivity or objective reality, but of truth. If we were to ask a member of the crowd if the person on whom they were all converging were in fact responsible for the crisis in the first place, if he actually deserved to be killed, he or she would surely say yes. But that is precisely the sort of question that no one will ask in the midst of a mimetic crisis. This question, if it is asked at all, will only be asked well after the crisis is over, when the answer can be safely rationalized without threatening the solidarity of the crowd.

The temptation to answer, "Yes, that person is indeed responsible for the whole thing" will be very strong for anyone involved at all in the mimetic unanimity that turns the victim into the solution to the crisis. But the fact that this answer is a temptation, and the desire to avoid the question in the first place, betray the fact that it is not so much a mistake as it is a lie. The objective reality of the victim as the one to blame is determined by the fact that everyone agrees to this lie. The ultimate *truth* of the victim as scapegoat, however, can only be viewed unambiguously from a perspective outside the mimetic unanimity of the crowd. Given this unanimity, this perspective can only be that of God. In the end, the objective reality of the victim as solution to the mimetic crisis, as the one *to blame*, corresponds exactly to the lie of the victim as the one who *should be blamed*. In other words, what is real in the eyes of the crowd is not necessarily what is true in the eyes of God.

If this is true, then we must make an important correction to the conclusion we came to in the last chapter. We have seen how the quantum wave function, which is a scientifically complete description of objective reality, is in fact a reflection of the interdividuality of the human observer. The uncertainty that lies at the heart of nature is not an inherent randomness of the physical object but the natural contingency of human mimetic relationships. In short, we cannot separate quantum mechanics from the human observer and his or her models. We therefore concluded (very reasonably) that quantum mechanics—that is, quantum mechanical behavior—effectively came into being with the birth of humanity.

We now see more clearly that this same uncertainty reflects a gap between objective reality and divine truth. As human beings we are necessarily interdividual, and a human observation has no choice but to be entangled with another's observation. But who that "other" is matters. One's observation can be modeled after the observing intellects of other humans or after the observing intellect of God. In the former case the observation is objectively real; in the latter case the observation is not only real but true in the fullest sense of the word. With God as the model of observation, the human and divine intellects converge on the one true form of the object, which is the form generated in the creative intellect of God. There is no room for uncertainty here, only alignment with the ultimate truth.

Any uncertainty in an observation can only come from a triangular relationship with other human models, as the result of a deviated transcendence. When modern physicists insist that quantum theory is only about objective reality, they are more correct than they realize, for that is precisely *all* it is about. The deviated transcendence that gives rise to quantum mechanical uncertainty is the basis for objective reality but not for divine truth. It gives rise to a reality that has replaced the one true model of the divine intellect with the superimposed models of the mob. The correlation of quantum mechanics to humanity, therefore, is more specifically the correlation to a humanity that has turned away from God. In other words, quantum mechanics came into being, not with the birth of humanity, but with its Fall.

In an effort to develop a theory of ethics based on Thomistic principles, Josef Pieper works within the context of the "real" world in relation to God but does not consider the possibility of relating to the real world *instead of* God. With a clear continuity in mind between the human connection to reality and reality's connection to God, Pieper states the following: "Evil ultimately proves to be an 'ontic' contradiction, a contradiction of being, something that opposes reality, that does not correspond to 'the thing.'"[14]

However, precisely because evil is a contradiction of being, it opposes *truth*, not necessarily reality. Satan, after all, is the father of lies. If the Bible tells us that he is also the "prince of this world," then presumably we are talking about the real world, the objective world in which we live, but one that suffers from a lack of God—that is, from a lack of divine truth. Satan is the prince of a world where truth corresponds to an objective reality based on the horizontal transcendence of our mimetic models, rather than the vertical transcendence of God—a human truth that serves as a kind of uncertain and unstable substitute for divine truth.

Without distinguishing between horizontal and vertical transcendence, there is no way to recognize the gap between objective reality and divine truth. Thus Pieper comes to the following conclusion: "'Objectivity,' if thereby we mean 'fidelity to being,' is the proper attitude of man."[15] But objectivity is not fidelity to being, it is fidelity to the mob. Fidelity to being is fidelity to truth, to the creative mind and will of God. It is therefore, I think, more correct to say that the proper attitude of man is simply what we have always been told—not objectivity but faith.

7
The Imitation of Truth

Michelangelo's famous statue of David in the Galleria dell'Accademia in Florence is comprised, as are all objects, of form and matter. According to Aristotle and Aquinas, the essence of the statue is in its form. Its matter preexisted the statue and will continue to exist after the statue is broken. The form, on the other hand, is that which sprang from the creative intellect of Michelangelo, and is what defines this statue as *the* famous statue of David.

In Delaware Park in Buffalo, New York, there is a full-size replica of Michelangelo's statue. This statue is made of different material than the original in Florence, but has (for the sake of argument) the same form. It is therefore "essentially" the same as the statue in Florence. An artist might say the statue in Buffalo "captures the essence" of Michelangelo's famous statue.

Of course, this does not mean that the statue in Buffalo *is* the statue in Florence. While the former shares the same essence and therefore derives its existence from the latter, it does not share the same matter and therefore does not share the same being. This is a basic ontological constraint: two objects cannot be the same if they are comprised of different matter, even if they have the same form. One would think the alternate situation—two objects with the same matter but different form—is not even worth mentioning. From the perspective of Scholastic philosophy (and classical physics), this situation is impossible. Two objects that have the same matter are necessarily the same object (they share the same being) and consequently must have the same form. At the quantum level, however—that is, at the level of the fundamental building blocks of physical reality—we find that the rules are not so strict. It turns out that a physical object can indeed be comprised of different forms, represented by the components of the wave function, despite being associated with the same matter.

Not sharing the same being means not sharing the same truth, as discussed in the last chapter. The statue of David in Buffalo is an imitation—it pretends to be something it isn't. It is, so to speak, not telling the truth about itself, for while it is a real statue of David, it is not *the* real statue of David originally created by Michelangelo. In fact, it would be more precise to say that the statue in Buffalo, while perfectly real, is not the *true* statue of David. This is not merely a statement about the essence of the statue but a judgment on the existence of it.

At the quantum level, this "lie" of objective reality is not merely a metaphor. As objectivity increases, uncertainty decreases but never reaches zero. Based as it is on the deviated transcendence of our mimetic models of observation, objectivity is always striving for that ontological independence found only in divine truth. In this sense all objective reality is ultimately an imitation of truth—all the more so because this imitation is indeed defined by the observer's mimetic relationships with his or her models.

We human beings cannot be human without being mimetic—all humans are interdividuals. It would seem, then, that objective reality, the entire world around us, bound as it is to the intersubjective basis of the human observer, is inevitably doomed to be a sham, an approximation of the truth at best. However, while we human beings are indeed bound to our models, it matters who our models actually are. It makes a difference whether our imitation is the fuel for rivalry or the inspiration for love.

Both objective reality and divine truth involve a kind of transcendence—the former horizontal (or deviated) and the latter vertical. If we are to close the gap between them, we must align these two cardinal vectors. We must point our deviated mimetic relationships toward the "true north" of vertical transcendence. Is this even possible? Can reality be converted to the truth, and do we mere human observers have the power to carry it out?

I believe the answer is yes, and the evidence for it is in a curious detail found in a few specific places in the Christian Gospels. It is a detail completely unexplained and mentioned only in passing, yet it stands out for its awkwardness, its out-of-place-ness. I'm speaking of the fact that Jesus was somehow not recognized even by his own disciples only a few days after his crucifixion and resurrection.

There are at least three passages that refer to this lack of recognition explicitly. The first is Luke 24:13–32, which recounts the two disciples' encounter on the road to Emmaus: "While they were talking and discussing, Jesus himself came near and went with them, but their eyes were kept from recognizing him." The second passage is John 20:11–18, which describes Mary Magdalene's encounter with Jesus in the tomb: "When she had said this, she turned around and saw Jesus standing there, but she did not know that it was Jesus." The third passage is John 21:1–14, which describes the encounter of seven disciples with Jesus at the Sea of Tiberias: "Just after daybreak, Jesus stood on the beach; but the disciples did not know that it was Jesus."

For the sake of brevity, I will refer to this difference in the appearance of Jesus, and the resulting lack of recognition by Mary and the disciples, as the "postresurrection transfiguration" of Jesus. This of course is not to be confused with the Transfiguration, in which Peter, James, and John see Jesus in dazzling white clothes talking with Elijah and Moses. The small risk is worth taking because, as we will see shortly, the word "transfiguration" most accurately describes the phenomenon that we want to discuss. Moreover, some biblical scholarship interprets the preresurrection Transfiguration as a kind of prefiguration of the resurrected body of Jesus. It is therefore appropriate to associate the two miracles via the same terminology.[1]

What is important to note is the physicality of this difference in appearance. It is not merely a spiritual transformation, or a psychological transference, but a physical transfiguration. And yet no one understands this to mean that Jesus "disguised" Himself by changing His physical features. The disciples did not fail to recognize Jesus because he was taller or had a different nose. The Gospels state only that the people looking at Him did not recognize Him. When they do finally recognize Him, there is no mention of His features changing back to their original form. It is rather something Jesus does or says that effectively opens their eyes.

The implication, then, is that the physical body of Jesus is somehow the same body He always had. It is not a new body regenerated from heaven, but the same body resurrected from the grave. The clearest demonstration of this is John 20:27, in which Jesus invites Thomas to touch the wounds from His crucifixion, as proof that it really is Jesus's body. How is it, then, that Jesus was not recognized even by the people who knew Him best? How could His body be the same, down to His wounds, yet different to the point of being unrecognizable?

To put the question a little differently, How can the physical body of Jesus have different forms, such that people do not recognize the "Jesus-ness" of Jesus, despite being comprised of the same matter? We see right away the parallel with the same strange characteristic of physical reality at the quantum level. If we restrict our discussion to a comparison of physical properties (as most physicists would prefer), this may seem no more than a coincidence or perhaps an amusing analogy. But if we dig a little deeper, we will find a common ground between the quantum level and the Christological level, and this common ground is the sphere of mimetic relationships.

The relationship of Jesus to the Father can be understood as an example of perfect mimesis, or perfectly positive mimesis, otherwise known as love. The Father is the perfect model for the Son, who gives Himself willingly and completely to the Father. But whereas mimetic rivalry leads to undifferentiation, the relationship between Jesus and the Father, completely free of rivalry, results in a perfectly clear differentiation. The Son and the Father are not doubles attempting to acquire each other's being, they are distinct persons sharing to the fullest extent each other's being. This is maximally mimetic to the point where Son and Father are "consubstantial," beyond even the most extreme consequences of mimetic undifferentiation. And yet there is no confusion: Jesus is not a model for the Father, the Father is the sole model for Jesus. In other words, Jesus's existence, aligned so completely with the will of the divine intellect, is one of perfect truth, and as such involves no quantum uncertainty.

The rest of us human beings, by contrast, live in a world of uncertainty, defined to some degree by the deviated transcendence of our triangular observations. This is the quantum mechanical nature of our objective reality, which minimizes uncertainty but never reaches the certainty of divine truth. For Jesus this gap between objective reality and divine truth is simply not there, and the Thomistic ideal actually applies—for Jesus, all that is is indeed true. Now, it is certainly nothing new to say that Jesus had a unique relationship with the Father. What is important to realize for our purposes is that this unique relationship between Jesus and His model implies a unique relationship between Jesus and physical reality.

Before the existence of human beings there was no mimetic desire, no distinction to be made between horizontal and vertical transcendence, and therefore no difference between reality and truth. They were the same thing. With the birth of humanity came the first intersubjective interdividual observers, and this now made it possible to distinguish between the reality of an object as it was observed and the truth of the object as it was created. But all these things—reality and truth, subjectivity and objectivity—while distinguishable in principle, need not have been misaligned with each other. Each one could very well have implied the others, much as Aquinas imagined. In biblical terms, we are talking about the prelapsarian Adam, whose only model was God and whose only physical reality was Paradise.

It is with the Fall of Adam that these elements must have begun to break apart and separate. As they did so, a new dimension opened up, a horizontal transcendence, that resulted in a kind of "transfiguration" of physical nature, which we now call "quantum mechanics." What Jesus did was to bring these elements back together again. Jesus has been understood as representing a *new* kind of human being: a "new Adam." He is, of course, uniquely perfect in His humanity. But this does not mean that he is a perfect example of our humanity, but rather a new example for humanity to follow. He is a human being who relates to the world in a fundamentally different way, characterized by a different subject-model relationship. So, if the Fall of the first Adam resulted in a transfiguration of physical nature, then it is only reasonable to suppose that the resurrection of the *new* Adam would result in a *new transfiguration* of physical nature.

Like the first transfiguration, this second one would most likely not involve a wholesale change in the laws of physics on some global scale. It would allow for differences in the behavior of nature in the presence of this new subjective observer, in relation to Him. For the rest of humanity, the world would continue to operate as before, as their relation to nature would be unchanged, determined by the same imperfect mimetic relationships.

But what would happen if an old observer were to observe this new human? By entering into a mimetic relationship with Jesus, another person would participate to some degree in the new triangular structure that transfigured physical reality in relation to Jesus. Now, in order to enter into a mimetic relationship with nature, one must first have had a mimetic relationship with a model. Even if this connection is an existential one stretching back to the collective origins of humanity, the first experience of mimesis is the relation with the Other. Therefore, the first experience of the new transfigured reality would have to be via an experience of Jesus. By "experience" we mean in this case the observation of physical reality. A person would essentially catch a first glimpse of this new physical reality in his or her observation of the physical reality of Jesus—that is, in Jesus's body. In other words, the first experience of the new reality opened up by the presence of Jesus would most likely be the observation of a physical transfiguration of Jesus's body.

If this is true, then we must address a rather obvious

question: If this transfiguration of Jesus's body is ultimately the result of His unique relationship with the Father, why does it only manifest itself *after* the resurrection? Didn't this relationship exist before the resurrection as well? Was Jesus not the perfect Son both before and after? The Bible offers nothing by way of an explanation, and there is nothing to suggest any difference in this Father-Son relationship, except for a small clue in John 20:17. Immediately after Mary recognizes Jesus in the tomb, He says to her, "Do not hold on to me, because I have not yet ascended to the Father."

Jesus makes no such comment or prohibition anywhere else in the Gospel. He clearly was not yet ascended to His Father before the resurrection, and yet there was apparently no issue with hugging Him, washing His feet, or helping Him up from the ground. Now this same condition is used as the basis for a new restriction pertaining to His physical body. So there is a difference, and moreover it has something to do with Jesus's relation to the Father. This is all we can glean from the Gospels, but it is enough to support what we have said thus far with regard to triangular observation. The transfiguration of Jesus is a transfiguration of physical nature that is ultimately the result of His unique relationship with the Father, which is unique in such a way as to produce this effect only after the resurrection.

What did this transfiguration, this new reality, look like? It is tempting to suggest that reversing the effects of the first transfiguration should bring us back to a reality defined by classical physics, which can be called a nonquantum reality. But this would not be quite accurate, because classical physics is not simply reality without quantum mechanics. What we know as classical physics is actually an approximation of quantum physics. It is what quantum mechanics looks like on a macroscopic scale. In other words, the classical physics with which we are familiar cannot exist without quantum physics. By contrast, the prelapsarian reality brought about by Jesus is the zero-uncertainty reality that points solely and consistently to the truth born in the divine intellect. We can describe this reality in logical and philosophical terms, but we have never actually seen it, at least not clearly. Whatever the transfigured reality of Jesus actually looked like, it was most likely not easily recognized.

And yet it could not have been *completely* unrecognizable, as its basis in truth provides a sort of model for quantum physics to imitate. We cannot know exactly what it looks like to fully close the gap between reality and truth, but we know our objective point of view peers in that direction. This allows for the possibility at least of recognition. Our objective reality can be compared to, say, a biographical play about Napoleon, in which the actor on stage imitates the true Napoleon as best he can. If the audience were to go back in time to the Napoleonic Wars, they would have a hard time recognizing Napoleon at first: he would look and speak somewhat differently than the stage actor did. But with the help of certain clues—the imposing personality, the habit of resting his hand inside his waistcoat—the audience may indeed eventually come to recognize him. So it must have been with the transfigured reality of Jesus, imitated as it was by the quantum reality people were used to seeing.

Consequently, while the hypothesis of triangular observation cannot give us a precise description of this new transfiguration of physical reality, it does allow us to make three basic statements about it. First, the initial experience of this new transfiguration was necessarily in the observation of the transfigured body of Jesus. Second, whatever this transfiguration actually looked like, it was most likely unrecognizable at first. Third, with some effort and help, it should have been possible to recognize it eventually.

This happens to be everything said by the Gospels themselves about the postresurrection transfiguration of Jesus. They say nothing about what Jesus actually looked like or why He should have looked different in the first place. We know only that it was the resurrected body of Jesus that was somehow the same and different, and different in such a way as to be unrecognizable at first, but eventually recognized. We therefore have arrived at a complete description of this postresurrection miracle, exactly as it is described in the Gospels, except that we have described it entirely in terms of the principles of triangular observation.

Science and Religion

We are used to hearing these days about the great debate between "science" and "religion." The degree to which this debate is serious or even valid is, of course, questionable. Nonetheless, there has always been a distinction between these two broad and ambiguous terms, although this distinction has always been somewhat dynamic, and the lines between the two camps have always been somewhat blurry. Beginning with the primitive terror of the sacred and moving through the sacrificial piety of the Hellenistic age from which Western science ultimately emerged, many would argue that the gap between science and religion progressively widened throughout the Enlightenment and the "Age of Reason," ultimately culminating in the modern mechanistic attitude that turns cynically away from all things human or divine.

As a general approach to the "problem" of mixing science and religion, this understanding of the situation has some truth to it. Our general attitude toward the physical world is certainly very different from that of our distant ancestors. For them the world of nature was saturated with the sacred. It was an object of fear and awe, something one shouldn't tamper with unless absolutely necessary, and then only if one followed the strictest rules. We moderns no longer see gods in the mountains or fairies in the forest. But while we have learned to trust nature, the more modern among us still seem to distrust the sacred as much as the ancients did. Every ancient sacred ritual was an attempt to avoid, appease, or control the sacred, and every modern appeal to a purely physical or mechanistic "objectivity" is essentially the same. Our ancestors turned from the sacred in fear and trembling, and we moderns turn away in haughty disdain. But we turn away just the same.

It turns out, in fact, that our most distant ancestors made similar distinctions between religion, which had a more explicitly social function, and other more pragmatic elements of their culture. They of course did not practice science in any modern sense of the word, but they did practice something eerily similar: magic. The anthropologist Bronislaw Malinowski has noted that the practice of magic in primitive cultures was not at all the romantic adventure of fantasy we usually associate with wizards and dragons. It was, in general, fairly prosaic and pragmatic, administered for mostly practical reasons according to very specific rules and formulas, much like the current practice of science. And like modern science, magic was always directed toward a physical object—the charm in which the power of magic was placed. If religion was the sacred in relation to the social or collective, magic was the sacred in relation to nature.

But perhaps because of primitive man's closer relationship with the sacred, he understood, in a primal intuitive way, that this connection to nature was not independent of the human person. He understood that if nature exhibited some strange force or will—a fickle inconsistency with one object, or a distant unseen sympathy with another—this will could not come from the object itself. While archaic magic shares much in common with modern science, the primitive magician was less confused about this than the modern scientist.

Magic was something that pertained to nature, but only insofar as nature pertains to man. "Magic is the quality of the thing, or rather, of the relation between man and the thing, for though never man-made, it is always made for man."[1] Like everything else in archaic culture, this understanding of nature was permeated with a fundamental fear of the sacred. But the basic intuition that nature does not merely exist but exists also in relation to the human is, to some degree, what I would like to rekindle in the present study.

The relationship between science and religion, or between the secular and the sacred in general, is of course another one of those big subjects that we cannot delve into here. My purpose in bringing it up is simply to acknowledge that this short monograph, which unabashedly blurs the lines between science, philosophy, metaphysics, and theology, will surely rub people the wrong way for precisely this reason. But we should understand that this vague discomfort that rubs people the wrong way has its roots in the sacred fear that rubbed people out on the sacrificial altars of our ancient ancestors. If today the tension between science and religion feels especially acute, this is certainly not because the gap between them is getting wider, which should only make us feel more at ease. The enormous growth and popularity of literature dealing with the mystical and metaphysical implications of quantum theory, much of it written by leading figures in the scientific community, reflects an overall impression that this gap is actually getting smaller. The strangest thing about the last hundred years is that we seem to have taken a quantum leap, not in the direction of a more mechanistic or objectively unambiguous universe, but into Schrödinger's "blurred model of reality," which blurs precisely those distinctions we only recently started taking for granted. Our understanding of the world has not merely gotten more complex or more sophisticated, it has gotten decidedly more mysterious.

Of course, this does not mean we should abandon ourselves to the deluge of mystical interpretations of quantum theory put forth over the last several decades. While many of these ideas may hold worthwhile insights or open valid questions, the philosophical and scientific communities are rightly suspicious of the facility with which these books have filled the New Age section of the bookstore.

This monograph will no doubt share the same shelf with some of those, and perhaps this is appropriate, or at least unavoidable. But it is worthwhile to note that the hypothesis of triangular observation presented here distinguishes itself from most interpretations of quantum theory in at least two important ways.

First, triangular observation does not suggest in any way that quantum mechanics is not real. This notion is explicitly or implicitly embedded in many "nonscientific" interpretations of quantum theory, which gives the impression that some sort of lie has been revealed, or some imposter has been unmasked. But we cannot so easily discount the uncanny accuracy with which quantum mechanics has described the world over the last century. We discussed earlier the modern scientist's stubborn denial of subjective reality and the humanity of the human observer, and we noted that this is ultimately an attempt to solve the problem by avoiding it altogether. The stubborn denial of the reality of quantum mechanics is just the other side of the same coin. Triangular observation not only acknowledges the reality of quantum mechanics, it recognizes its fundamental characteristics in the strong realism of thinkers like Thomas Nagel and Thomas Aquinas.

Second, while triangular observation insists on realism, it does not simplistically equate reality with physical reality. Here we are in agreement with Nagel (and many others) that if physics is to be a complete model of the real world, as it claims, then it must include *all* aspects of the real world. This includes what we call subjective and objective realities, as well as what we call human and divine realities. These aspects of the real world are not independent of each other, and taking their interrelationships into account allows us to broaden the scope of science without losing our grip on reality in general. It allows us to explain something like the postresurrection transfiguration of Jesus without explaining it away. Consequently, when we explain something "scientifically," this means something more than reducing everything to a set of physical principles. On the contrary, the concept of triangular observation *elevates* physics above the opaque obstacles of physicalism and opens its view to all things related to the natural world, including our place as human observers within it. It was this same holistic view of science that inspired the noted physicist John Wheeler to say the following: "No theory of physics that deals only with physics will ever explain physics. I believe that as we go on trying to understand the universe, we are at the same time trying to understand man."[2]

In the dualistic model of the world that divides everything into inside and outside, subjective and objective, it is understandable that "nonscientific" interpretations of quantum theory would be treated with some skepticism. Most interpretations in fact set themselves up for this cold reception by following the same dualistic attitude. Some aspect of the nonphysical world (consciousness, belief, oneness, etc.) effectively imposes itself on the physical world, like a ghost imposing its unwelcome spell on the solid walls of a house. Once this piece of the nonphysical world has crossed over into the realm of the physical, we are expected to treat it as another piece (perhaps a special piece) of the physical world and analyze it with the appropriate methods and attitudes of the physicist. But the physicist wants to ask why this particular piece of the nonphysical world should cross over into his domain and be converted to physicalism, and the philosopher or theologian wants to ask the same thing. There is something unconvincingly arbitrary about the whole thing. What is so special about, say, consciousness? Why consciousness and not conscience? Why belief states and not statements of faith? If the physical world is indeed related ontologically to the human observer, at the level of the physical laws themselves, then it cannot be related merely to some piece of humanity that can be conveniently treated or analyzed as part of the nonhuman world, as if its humanity were simply an inappropriate starting point. If it is the human observer that matters, then the physical world must be related to that which makes us specifically human—that is, to that which not merely relates us to but *distinguishes* us from the rest of the world.

What is it that makes us specifically human? This is yet another one of those big questions that we keep running into. The fact that we can't avoid these questions only confirms the basic point we are trying to make—namely, that physics cannot continue to avoid the human or the divine by hiding behind the claim that it only pertains to the physical world. By doing so it only reduces the scope of physics to something less than the real world, whose claim to objectivity relies on the multiplicity of observers that agree on the state of this world. This multiplicity, in turn, derives from the sphere of mimetic models that defines the observer as human. "Men, not man, inhabit this planet," Arendt reminds us.

To the question of what makes us specifically human, therefore, we can offer a short answer here. It is the same thing that makes us objectively real—our relationships to those observers around us—with the crucial distinction that for us humans those relationships are fundamentally mimetic. What distinguishes us humans from other physical objects, including other measurement devices, is not that our reality is somehow irrelevant for the physical world, but that our relationship to this world is such that it transcends the mere subject-object relationship currently envisioned by the physicist. The physical world, in its relation to the human observer, is related to that observer's mimetic models, both human and divine.

If one wants to draw a line between modern science and religion, then this line cannot be drawn between the physical and the human—if the observer's models are human, the triangular structure of observation makes these components of the world inseparable. Neither can the line be drawn between the physical and the divine—if the observer's model is divine, following the logic of Aquinas, these two realities, the created object and the creative intellect, are similarly inseparable. The line must therefore be drawn, not within a particular mimetic triangle, but *between* different triangles. More specifically, the difference between modern science and religion, between the study of the objectively real world and the study of the divinely created world, is the difference between the world as defined by an observer's human models and the world as defined by an observer's divine model. In other words, the line between science and religion is the line between objective reality and divine truth. The physicist is justified in claiming that his field deals with what is real and observable, but he must ultimately look to God to know if his observations are true. This is, after all, what every scientist of good conscience knows intuitively, regardless of his or her specific understanding of God. And it is the solid intuition of these good people that allows science to progress, faithfully, in the right direction.

Notes

Introduction

1. Stephen Hawking and Leonard Mlodinow, *The Grand Design: New Answers to the Ultimate Questions of Life* (Bantam Press, 2010). This is, of course, a rather offhand and often-quoted example of scientists' attitude toward philosophy, made by one of the more famous people in the field. But there are many other examples, and I will mention a few of them throughout the course of the monograph.

2. The well-known "Schrödinger's cat" thought experiment, which was the first and clearest expression of the strangeness of quantum theory, will be discussed in detail in chapter 1.

Chapter 1. The Collapse of the Rational World

1. The experiment measured the photoelectric effect, in which electrons are ejected from a sample of material by illuminating it with a beam of light. One would expect the number of electrons ejected to be proportional to the intensity and wavelength of the incident light. But it turns out that beyond a certain wavelength, no electrons are ejected at all.

2. Erwin Schrödinger, "The Present Situation in Quantum Mechanics," translated by John D. Trimmer, *Proceedings of the American Philosophical Society* 124 (1980): 323–38.

3. Ibid.

4. Henry P. Stapp, *Mindful Universe: Quantum Mechanics and the Participating Observer* (New York: Springer, 2011).

5. In fact, the most appropriate term is "transcendent," which more clearly expresses the notion of an observer observing himself. However, this term is even more overloaded and potentially misleading than the term "consciousness." Later in our study we will reach the right point in which to define these concepts more precisely. But for now we will stay with a terminology more familiar in discussions of the

measurement problem.

6. The assumption here is that humans are distinguished from other animals by this unique quality of self-awareness, as opposed to the instinctual consciousness of other animals. This distinction, however, is not important for the development of the ideas proposed here. Any reader who would include certain higher mammals within the subset of self-aware beings is free to do so.

7. For example, in Carlo Rovelli's paper on relational quantum mechanics, his first hypothesis states that all systems (including all observers) are equivalent, and there is nothing that makes any one type of system (such as a human being) special compared to others. He then states with admirable honesty, "Of course, I have no proof of hypothesis 1, only plausibility arguments. I am suspicious toward attempts to introduce special non-quantum and not-yet-understood new physics, in order to alleviate the strangeness of quantum mechanics."

8. Martin Heidegger, "Modern Science, Metaphysics and Mathematics," *Basic Writings*, rev. ed. (San Francisco: HarperSanFrancisco, 1993), 291.

9. I will discuss some of these in more detail later in the monograph.

10. There are several such systems that have been used to describe this famous thought experiment. For example, electron-positron pairs have opposite spins, and photons can have opposite polarizations. Einstein, Podolsky and Rosen used position and momentum: if one is known with 100 percent certainty, the other is not known at all. *Physics Review* 47, no. 10 (1935): 777–80.

Chapter 2. Modern Observation through Medieval Eyes

1. Thomas Nagel, *The View from Nowhere* (New York: Oxford University Press, 1986), 20.

2. Ibid., 127.

3. See, for example, Jerome R. Busemeyer and Peter D. Bruza, *Quantum Models of Cognition and Decision* (Cambridge: Cambridge University Press, 2012).

4. Josef Pieper, *Living the Truth* (San Francisco: Ignatius Press, 1989), 55.

5. Thomas Aquinas, *Summa Theologica 1*, 14, 1.

6. This idea will seem even more appropriate if we consider

the notion of the "creative intellect" as opposed to the "receptive intellect," which will be discussed in chapter 6.

7. Pieper, *Living the Truth*, 55.

8. Ibid.

9. Roger P. Paul, "Subjectivist-Observing and Objectivist-Participant Perspectives on the World: Kant, Aquinas and Quantum Mechanics," *Theology and Science* 4, no. 2 (2006): 165.

Chapter 3. The Interdividual Observer

1. Fyodor Dostoyevsky, *Notes from Underground*, trans. Jessie Coulson (London: Penguin, 1972), 1.

2. Albert Camus, *The Stranger*, trans. Matthew Ward (New York: Vintage, 1988), 59.

3. The book in which Girard first introduces and begins developing his ideas on mimetic desire is *Deceit, Desire and the Novel*. For a more complete discussion of mimetic theory and its main implications in psychology, anthropology, philosophy and theology, read *Things Hidden since the Foundation of the World*, trans. Stephen Bann and Michael Metteer (Stanford, CA: Stanford

University Press, 1987). For an excellent study of how mimetic theory is applied to both ancient and modern cultures, read *Violence Unveiled* by Gil Bailie (New York: Crossroad, 1995).

4. In fact we do see examples of a basic "acquisitive mimesis" among some higher animals, and this is an important aspect of Girard's anthropological hypothesis on the origin of humanity. But the present effort is concerned more with the nature of human desire than its origins.

5. It is important to distinguish, however, between the abstract concept of "humanity" and the very real people and relationships that make up human culture.

6. Bernard d'Espagnat, *Veiled Reality: An Analysis of Present-Day Quantum Mechanical Concepts* (Boulder, CO: Westview Press, 2003).

7. I will discuss some of this work in more detail in the next chapter.

Chapter 4. People and Things Hiding in Plain Sight

1. David Bloor, "Anti-Latour," *Studies in History, Philosophy & Science* 30, no. 1 (1999): 87.

2. Bruno Latour and Steve Woolgar, *Laboratory Life: The Construction of Scientific Facts* (Princeton: Princeton University Press, 1979), 32.

3. Bloor, "Anti-Latour," 109.

4. Ibid., 107–8.

5. Ibid., 109.

6. Thomas S. Kuhn, *The Structure of Scientific Revolutions* (Chicago: University of Chicago Press, 2012), 11.

7. Niels Bohr, *Essays 1932–1957 on Atomic Physics and Human Knowledge* (Woodbridge, CT: Ox Bow Press, 1958), 39.

8. Niels Bohr, *Essays 1958–1962 on Atomic Physics and Human Knowledge* (Woodbridge, CT: Ox Bow Press, 1963), 7.

9. It is interesting to note that many-worlds theory was not taken very seriously by the scientific community for a long time because of its rather far-fetched metaphysical implications. In recent years, however, it has gained a larger following among theoretical physicists for the simple reasons that (*a*) it is mathematically consistent and (*b*) no one is coming up with anything better.

10. David Albert and Barry Loewer, "Interpreting the Many Worlds Interpretation," *Synthese* 77 (1988): 203.

11. Ibid., 211.

12. Carlo Rovelli, "Relational Quantum Mechanics," *International Journal of Theoretical Physics* 35, no. 8 (1996): 1637–78.

13. Roger Scruton, *An Intelligent Person's Guide to Philosophy* (New York: Penguin, 1996), 47.

14. Stephen L. Gardner, *Myths of Freedom* (Westport, CT: Greenwood Press, 1998), 79.

15. Scruton, *An Intelligent Person's Guide*, 49–50.

16. René Girard, *Deceit, Desire, and the Novel: Self and Other in Literary Structure*, trans. Yvonne Freccero (Baltimore: Johns Hopkins University Press, 1966), 16.

Chapter 5. The Roots of Reality

1. Niels Bohr, *Essays 1932–1957 on Atomic Physics and Human Knowledge* (Woodbridge, CT: Ox Bow Press, 1958), 90–91.

2. Thomas Nagel, *The View from Nowhere* (New York: Oxford University Press, 1986), 21.

3. Albert Einstein, *Relativity*, trans. Robert W. Lawson (New York: Crown, 1961), 140.

4. Ibid., 141.

5. Nagel, *The View from Nowhere*, 38.

6. Ibid., 81.

7. Ibid., 126.

8. Hannah Arendt, *The Life of the Mind* (New York: Harcourt, 1978), 49.

9. Nagel, *The View from Nowhere*, 125.

10. Luke 23:34.

11. Arendt, *Life of the Mind*, 25.

12. Eric Gans, *Signs of Paradox: Irony, Resentment, and Other Mimetic Structures* (Stanford: Stanford University Press, 1997).

13. But then, of course, we are thrown back into the world of problems and paradoxes described in chapter 1.

Chapter 6. From Reality to Truth

1. Werner Heisenberg, *Physics and Philosophy* (New York: HarperCollins, 2007), 160.

2. Wolfgang Smith, *The Quantum Enigma: Finding the*

Hidden Key (Hillsday, NY: Sophia Perennis, 2005), 85.

3. Ibid., 145.

4. Gerard Verschuuren, *Aquinas and Modern Science: A New Synthesis of Faith and Reason* (Kettering, OH: Angelico Press, 2016).

5. Josef Pieper, *Living the Truth*, trans. Lothar Krauth and Stella Lange (San Francisco: Ignatius Press, 1989), 55.

6. Thomas Aquinas, *Quaestiones Disputatae de Veritate*, Question 3, Article 3.

7. Ibid., Question 1, Article 3.

8. Ibid., Question 1, Article 2.

9. Ibid., Question 1, Article 3.

10. Ibid., Question 1, Article 10.

11. Ibid., Question 1, Article 1.

12. Ibid., Question 1, Article 2.

13. Someone who subscribes to a relativist philosophy, in which truth may be different for different people, may object to the notion of a single truth, or a single form corresponding to the truth. But this concept of relative truth is really only something closer to what I am calling "objective reality." There is still a philosophical distinction

to be made between an object that has the potential to be multiple things and an object that has the potential to be only what it is. The former is what I, following quantum theory, call "objective reality." The latter is what I, following Aquinas, call "divine truth."

14. Pieper, *Living the Truth*, 79.

15. Ibid., 80.

Chapter 7. The Imitation of Truth

1. My thanks to James Williams for bringing this important detail to my attention.

Epilogue. Science and Religion

1. Bronislaw Malinowski, *Magic, Science and Religion and Other Essays* (n.p.: Read Books, 2014), 75.

2. John Archibald Wheeler, *The Intellectual Digest*, June 1973.

Index